Current Topics in Microbiology 123 and Immunology

Editors

Retroviruses 4

Edited by P.K. Vogt and H. Koprowski

With 23 Figures

Springer-Verlag
Berlin Heidelberg NewYork Tokyo

Professor Dr. PETER K. VOGT
University of Southern California
School of Medicine
Department of Microbiology
2025 Zonal Avenue HMR 401
Los Angeles, CA 90033, USA

Professor Dr. HILARY KOPROWSKI
The Wistar Institute
36th Street at Spruce
Philadelphia, PA 19104, USA

ISBN 3-540-15947-9 Springer-Verlag Berlin Heidelberg New York Tokyo
ISBN 0-387-15947-9 Springer-Verlag New York Heidelberg Berlin Tokyo

Library of Congress Catalog Card Number 15-12910
Printed in Germany.

Typesetting, printing and bookbinding:
Universitätsdruckerei H. Stürtz AG, Würzburg
2123/3130-543210

Table of Contents

J.S. Brugge: Interaction of the Rous Sarcoma Virus Protein pp60^{src} with the Cellular Proteins pp50 and pp90. With 4 Figures 1

D.W. Stacey: Microinjection Studies of Retroviral Polynucleotides. With 9 Figures 23

B.M. Sefton: The Viral Tyrosine Protein Kinases. With 4 Figures 39

C. Van Beveren and I.M. Verma: Homology Among Oncogenes. With 6 Figures 73

Indexed in Current Contents

List of Contributors

BRUGGE, J.S., Department of Microbiology, State University of New York at Stony Brook, Stony Brook, NY 11794, USA

SEFTON, B.M., Molecular Biology and Virology Laboratory, The Salk Institute, P.O. Box 85800, San Diego, CA 92138, USA

STACEY, D.W., Department of Cell Biology, Roche Institute of Molecular Biology, Roche Research Center, Nutley, NJ 07110, USA

VAN BEVEREN, C., Molecular Biology and Virology Laboratory, Salk Institute for Biological Studies, P.O. Box 85800, San Diego, CA 92138-9216, USA

VERMA, I.M., Molecular Biology and Virology Laboratory, Salk Institute for Biological Studies, P.O. Box 85800, San Diego, CA 92138-9216, USA

Interaction of the Rous Sarcoma Virus Protein pp60^{src} with the Cellular Proteins pp50 and pp90

J.S. Brugge

1 Introduction 1
2 Identification of the Complex 2
3 Interaction of pp50 and pp90 with Other Oncogene Products 5
4 Specificity of the Interaction Between pp60^{src}, pp50, and pp90 5
4.1 pp50 and pp90 Bind to Newly Synthesized Molecules of pp60^{src} 5
4.2 pp90 and pp50 are Complexed with pp60^{src} Molecules Which Are Not Associated with the Plasma Membrane 6
4.3 pp60^{src} Associated with pp90 and pp50 Does Not Contain Phosphotyrosine 6
5 Protein Kinase Activity of pp60^{src} Bound to pp50 and pp90 7
6 Characterization of the pp90 and pp50 Proteins 7
6.1 pp90 7
6.2 pp50 9
7 Sites of pp60^{src} Which Interact with pp90 and pp50 11
7.1 Analyses of Viruses Carrying Mutations Within the *src* Gene 11
7.2 Analyses of the Complex Using Antibodies to Specific Regions of pp60^{src} 11
8 Model for the Interaction Between pp50 and pp90 12
9 Possible Functional Roles of the Interaction Between pp50, pp60^{src}, and pp90 13
9.1 Transport of pp60^{src} to the Plasma Membrane 14
9.2 Attachment of Myristate to pp60^{src} 15
9.3 Phosphorylation of pp50 16
9.4 Regulation of the Phosphotransferase Activity of pp60^{src} 17
10 Interaction of pp90 and pp50 with Nonviral Proteins 18
11 Future Directions 19
References 20

1 Introduction

Oncogenic retroviruses cause multiple and profound alterations in the morphology, metabolism, and growth control of cells. In most retrovirus infected cells, all of these complex changes in the cellular phenotype are mediated by a single virus-encoded gene product, referred to as the *transforming protein*. Rous sarcoma virus (RSV) has proven to be an ideal system for the analysis of events which occur following oncogenic transformation by retroviruses. RSV induces rapid sarcoma production after injection of virus in vivo and efficient and rapid transformation of cells in culture (Hanafusa 1977). The transforming gene

Department of Microbiology, State University of New York at Stony Brook, Stony Brook, NY 11794, USA

Current Topics in Microbiology and Immunology, Vol. 123

of RSV encodes a protein of M_r 60000 which has been designated pp60^{src} (BRUGGE and ERIKSON 1977; PURCHIO et al. 1978). Genetic studies of viruses encoding mutant *src* gene products which induce a partially transformed phenotype suggest that interactions between pp60^{src} and multiple cellular targets are required to elicit a fully transformed phenotype (review, SEFTON and HUNTER 1984). This review will discuss one such interaction between pp60^{src} and host cell proteins. This interaction occurs between newly synthesized molecules of pp60^{src} and two cellular proteins of M_r 90000 (pp90) and 50000 (pp50). The kinetics and localization of this interaction suggest that the cellular pp90 and pp50 proteins are involved in some aspect of the processing of pp60^{src} before it reaches its residence in the plasma membrane (COURTNEIDGE and BISHOP 1982; BRUGGE et al. 1983). pp90 and pp50 have also been shown to associate with many retrovirus-encoded transforming proteins other than pp60^{src} (LIPSICH et al. 1982; ADKINS et al. 1982). This suggests that pp50 and pp90 may play a common role in the events which take place after transformation by at least one class of retrovirus transforming proteins.

The RSV-transforming protein has been shown to possess a tyrosine-specific phosphotransferase activity (COLLETT and ERIKSON 1978; LEVINSON et al. 1978; HUNTER and SEFTON 1980). All of the other retrovirus transforming proteins which interact with pp50 and pp90 also possess this enzymatic activity and share structural homology with pp60^{src}. This review will summarize the information which has accumulated on the nature of this complex and its protein components and discuss the possible functions of the interaction between pp50 and pp90 and viral transforming proteins which possess tyrosine-specific protein kinase activity.

2 Identification of the Complex

All of the initial experiments performed to characterize the pp60^{src} protein were carried out using an antisera obtained from rabbits bearing tumors induced by RSV. This antisera was not monospecific for pp60^{src} and consequently precipitated multiple protein species other than pp60^{src}. The immunoprecipitation of most of these proteins could be blocked by preabsorption of the antiserum with the RSV structural proteins, indicating that these proteins were related to viral structural proteins expressed in the tumor cells. However, two proteins of M_r 90000 and M_r 50000 were consistently precipitated from RSV-transformed chicken cells by the preabsorbed serum from tumor-bearing rabbits (TBR serum) (see Fig. 1 and SEFTON et al. 1978; BRUGGE et al. 1981; OPPERMANN et al. 1981b). Three possible explanations for the precipitation of these proteins by TBR serum were the following: (a) The pp50 and pp90 proteins were structurally related to pp60 as precursors or cleavage products; (b) TBR serum contained unique antibody molecules which recognized pp90 and pp50; (c) pp50 and pp90 were associated in a protein complex with pp60 and thus coprecipitated with antibody to pp60.

Possibility (a) was ruled out by analysis of peptides derived by partial or complete digestion of pp50, and pp60, and pp90 using a variety of proteolytic

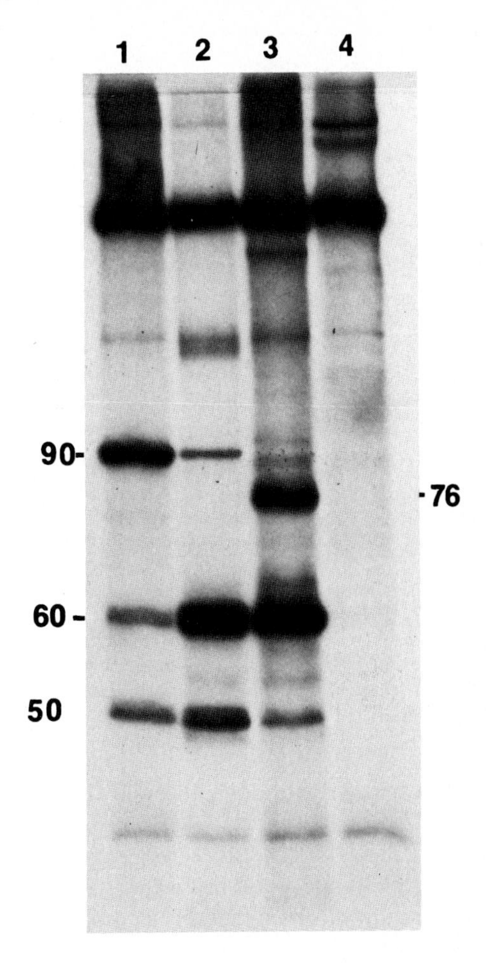

Fig. 1. Immunoprecipitation of pp60, pp50, and pp90 from ^{32}P-labeled RSV transformed chicken cells. RSV (Schmidt-Ruppin subgroup A) transformed cells were labeled with ^{32}P 4 h. Cell lysates were prepared, and the proteins were immunoprecipitated by the antisera below and analyzed on 7.5% SDS-polyacrylamide gels as described (BRUGGE and ERIKSON 1977). *Lane* 1, monoclonal antibody to pp90; *lane* 2, monoclonal antibody to pp60^{src}; *lane* 3, tumor bearing pp60^{src}, rabbit (TBR) serum; *lane* 4, control rabbit serum. Pr76 is the viral *gag*-gene translation product

enzymes (SEFTON et al. 1978; BRUGGE et al. 1981; OPPERMANN et al. 1981b). These studies did not reveal any structural similarities between the three proteins. Possibility (b) was ruled out because pp50 and pp90 were not precipitated from uninfected cells or cells infected with viruses containing large deletions in the *src* genes (BRUGGE et al. 1981).

Support for the last alternative, that pp50 and pp90 were associated in a complex with pp60, was obtained from several lines of evidence: (a) Sedimentation analysis of RSV-transformed cell lysates indicated that pp60 existed in two forms, the majority of the protein sedimented as a monomer 60000-dalton protein while a small percentage sedimented more rapidly in glycerol gradients (Fig. 2; BRUGGE et al. 1981). The pp50 and pp90 proteins were found to cosediment and coprecipitate with this faster sedimenting form of pp60. (b) Polyclonal antiserum prepared against pp90 immunoprecipitated pp60 and pp50 from the same gradient fractions which allowed precipitation of pp60, pp50, and pp90 using TBR serum (BRUGGE et al. 1981). (c) Monoclonal antibodies directed against pp90 coprecipitated pp60 and pp50 from RSV-transformed cells and

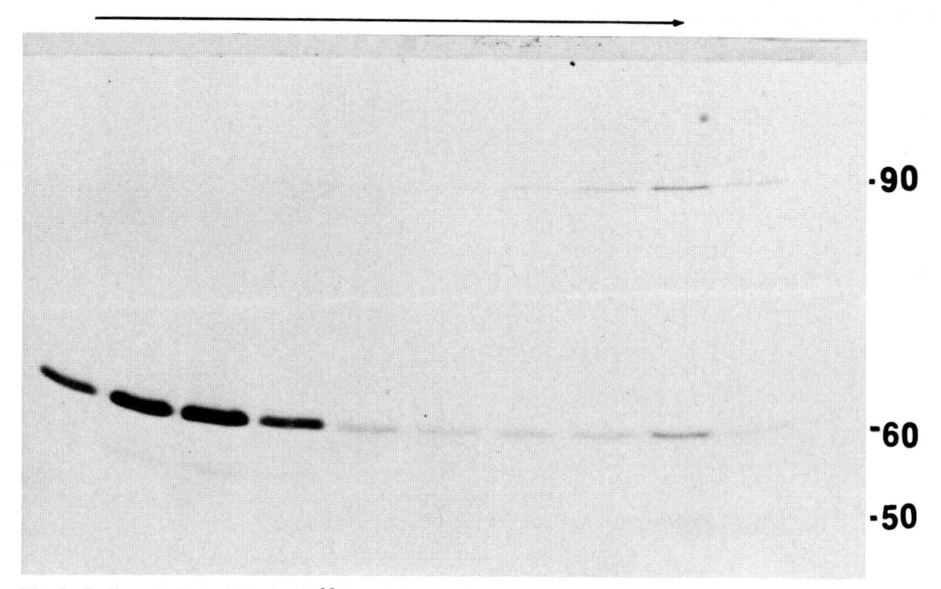

Fig. 2. Sedimentation analysis of ^{35}S-methionine labeled RSV-transformed chicken-cell lysates. SR-RSV (subgroup A) transformed chicken cells were labeled with ^{35}S-methionine for 24 h. A cell lysate was prepared and sedimented on a 10%–30% glycerol gradient as described (BRUGGE et al. 1981). Alternating gradient fractions were immunoprecipitated with antibody against pp60^{src}. Sedimentation was from *left* to *right*

monoclonal antibodies to pp60 (LIPSICH et al. 1983) also precipitated pp90 and pp50 (Fig. 1).

Taken together, these data provide strong evidence that pp60 is associated with pp90 and pp50 in lysates from RSV-transformed cells. Further characterization of this complex revealed that the interaction between these three proteins is very stable in vitro. The pp50:pp60^{src}:pp90 complex is not dissociated by incubation with high salt (2*M* NaCl, reducing agents, or metal-chelating agents); however, long-term incubation in the absence of salt dissociates the protein complex (BRUGGE et al. 1981).

The pp50 and pp90 proteins were detectable in pp60^{src} immunoprecipitates of cells infected with all nondefective strains of RSV; however, there were slight variations between virus strains in the levels of pp90 and pp50 precipitation (BRUGGE et al. 1983). Cells infected with mutant viruses displaying a temperature-dependent transformed phenotype showed elevated levels of pp50 and pp90 when incubated under either permissive or nonpermissive conditions with the highest levels being detected at the nonpermissive temperature (BRUGGE et al. 1981, 1983). Sedimentation analysis of lysates from cells infected with these mutant viruses indicated that the majority of the *src* protein was complexed with pp50 and pp90 at the nonpermissive temperature. These results indicate that mutations in the *src* gene which affect the transforming activity of pp60 also affect the binding of pp50 and pp90. This property was found in cells infected with a variety of ts mutants isolated using different methods of mutagenesis.

3 Interaction of pp50 and pp90 with Other Oncogene Products

Many of the retrovirus-encoded oncogene products have been shown to carry an associated tyrosine-specific protein kinase activity. pp50 and pp90 have been shown to co-immunoprecipitate with many of these transforming proteins including the *fps* gene products from Fuginami (LIPSICH et al. 1982) and PRCII sarcoma virus (ADKINS et al. 1982), the *yes* gene product from Yamaguchi 73 sarcoma virus (LIPSICH et al. 1982), and the *fes* gene product from Snyder-Theilen feline sarcoma virus (ZIEMIECKI, unpublished results). It has also been observed that mutant viruses carrying ts defects in the *fps* (S. MARTIN and T. PAWSON, unpublished results) and *fes* gene products (H. SNYDER, unpublished results) show elevated levels of pp50 and pp90 binding. MATHEY-PREVOT et al. (1984) have isolated a revertant line of Fuginami virus-transformed cells in which the majority of the p130^{fps} protein is associated in a complex with pp50 and pp90. This evidence indicates that several unique tyrosine-specific protein kinase transforming proteins bind to the same cellular proteins, and suggests that these proteins may play a common role in their interaction with all tyrosine-kinase transforming proteins. It is clear from immunoprecipitation experiments using monoclonal antibodies to pp90 that the association with these proteins is specific and not merely due to nonspecific interactions with protein present in cell lysates.

4 Specificity of the Interaction Between pp60^{src}, pp50, and pp90

The sedimentation analysis of pp60^{src} on glycerol gradients revealed that only a small percentage of pp60^{src} was bound to pp50 and pp90. These complex-associated pp60^{src} molecules could represent either a specific population which enter a nondissociable, dead-end complex with pp50 and pp90 or pp60^{src} molecules which were associated in a short-lived complex during some phase of their cellular lifetime. The evidence described below favors the latter possibility, that the interaction between pp90, pp50, and pp60^{src} is transient and involves all molecules of newly synthesized pp60^{src}.

4.1 pp50 and pp90 Bind to Newly Synthesized Molecules of pp60^{src}

The first hint that pp50 and pp90 bind to newly synthesized pp60^{src} molecules was the evidence that the percentage of radiolabeled pp60^{src} bound to pp50 and pp90 varied with the length of incorporation of ^{35}S-methionine into RSV-transformed cells. Analysis of cells labeled for short periods (2–5 min) indicated that more than 90% of radiolabeled *src* was associated with pp50 and pp90. The percentage of the radiolabeled form of complex-associated pp60^{src} decreased with increasing labeling periods until a steady-state percentage of 1%–5% was reached in 10-h labeling periods (BRUGGE et al. 1983). This suggested that newly synthesized molecules of pp60^{src} preferentially bind to pp90 and pp50. This possibility was further examined using pulse-chase experiments. The majority of pp60^{src} molecules labeled during a 15-min pulse was associated with pp50 and pp90. Within 30 min after removal of ^{35}S-methionine, approximately 70% of the radiolabeled pp60^{src} sedimented as a monomer. After 180 min, 95% of pp60^{src} dissociated from the complex and sedimented as a 60K protein (BRUGGE et al. 1983). This suggests that the complex has a half-life of approximately 15 min in cells transformed by the SR (subgroup A) strain of RSV. The complex was found to turn over more rapidly in cells transformed by the SR (subgroup D) strain of RSV.

In these experiments, it was noted that newly synthesized pp60^{src} did not bind to newly synthesized pp90 and pp50, since radiolabeled pp90 and pp50 were not coprecipitated with *src* molecules in pulse-labeled cells, but could be detected long periods after removal of ^{35}S-methionine. The detection of radiolabeled pp90 and pp50 that coprecipitates with pp60 during these chase periods indicates that molecules of pp90 and pp50 that are labeled during the 15-min pulse bind to unlabeled, newly synthesized *src* after the ^{35}S-methionine has been removed.

4.2 pp90 and pp50 are Complexed with pp60^{src} Molecules Which Are Not Associated with the Plasma Membrane

COURTNEIDGE et al. (1980) found that newly synthesized pp60^{src} is found in the soluble fraction of cells which are fractionated by homogenization in hypotonic buffers. Within 5–15 min after synthesis, pp60^{src} was shown to associate with the plasma membrane. Therefore, the time course of detection of pp60^{src} in the soluble portion of the cell shows the same pattern as the binding of pp60^{src} to pp50 and pp90; that is, newly synthesized pp60^{src} enters a complex with pp50 and pp90 which has a half-life of 10–15 min. Fractionation of cells by the homogenization protocol of COURTNEIDGE et al. (1980) or by extraction of "soluble" cellular proteins with mild detergent in a cytoskeleton stabilizing buffer was found to separate the monomer form of pp60^{src} from the pp50:pp90-bound form of pp60^{src}, with the complexed form being exclusively associated with the soluble cell fraction. The majority of unbound pp60^{src} was found in the particulate cell fraction (BRUGGE et al. 1983).

4.3 pp60^{src} Associated with pp90 and pp50 Does Not Contain Phosphotyrosine

pp60^{src} has been shown to contain at least three sites of phosphorylation: two serine sites within the amino half of the molecule, and one tyrosine site at residue 416 within the carboxyl half of pp60^{src} (SMART et al. 1981; CROSS and HANAFUSA 1983). The tyrosine phosphorylation is believed to be mediated by autophosphorylation although this has not been rigidly demonstrated. When the complex bound form of pp60^{src} was isolated by either fractionation on a glycerol gradient, precipitation with antibody to pp90, or more crudely by fractionation of soluble pp60^{src}, no phosphorylation was detected in the carboxyl half of pp60^{src} (BRUGGE et al. 1981, 1983). This suggested that either pp90 and pp50 selectively bind to molecules of pp60^{src} which are not phosphorylated at tyrosine 416 or that pp50 and pp90 association prevented autophosphorylation after phosphate turnover at this residue. BRUGGE et al. (1983) attempted to address this question by examination of complex-bound pp60^{src} obtained from cells after 15 min incorporation with $^{32}P_i$. No phosphorylation of the COOH-half of complex-bound pp60^{src} was detectable under these conditions. This experiment supports the possibility that pp90 and pp50 prevent phosphorylation on residue 416; however, it does not rule out other explanations. Since other viral transforming proteins were not found to be deficient in tyrosine phosphorylation when bound to pp50 and pp90, the significance of this specificity for serine-phosphorylated pp60$^{v\text{-}src}$ is not clear (YONEMOTO, LIPSICH and BRUGGE, unpublished results; ZIEMIECKI, unpublished results). pp90 and pp50 might bind to pp60^{src} either during or shortly after translation and block autophosphorylation.

5 Protein Kinase Activity of pp60^{src} Bound to pp50 and pp90

Upon first identification of the pp50 and pp90 association with pp60^{src}, it was of great interest to determine whether the pp60^{src}-specific phosphotransferase activity was intrinsic to pp50 or pp90. This was ruled out by glycerol gradient sedimentation analysis of pp60^{src}-specific phosphorylation of TBR-IgG. All of the enzymatic activity was found associated with the monomer form of pp60^{src}. These results also suggested that pp90 and pp50 binding might inhibit the enzymatic activity of pp60^{src}, since no activity was found in the region of the gradient where the protein complex sedimented. This result was also supported by the absence of protein kinase activity in immunoprecipitates prepared with monoclonal antibody directed against pp90. In this assay, phosphorylation of the exogenous substrate, casein, was examined (YONEMOTO and BRUGGE, unpublished results). In contrast, KRUEGER and coworkers and SEFTON have been able to detect low levels of TBR phosphorylation by the complex-bound *src* protein using either glycerol gradients containing NP40 instead of triton, sodium deoxycholate, and sodium dodecyl sulfate in the absence of EDTA and EGTA (B. SEFTON, unpublished results), or milder washing conditions during immunoprecipitation (KRUEGER et al. 1984).

The experiments described in the above sections clearly defined the time and place of the interaction between pp60^{src} and the cellular pp50 and pp90 proteins.

6 Characterization of the pp90 and pp50 Proteins

6.1 pp90

pp90 is a relatively abundant cellular protein, representing approximately 0.5% of the total cellular protein. This protein fractionates exclusively as a soluble,

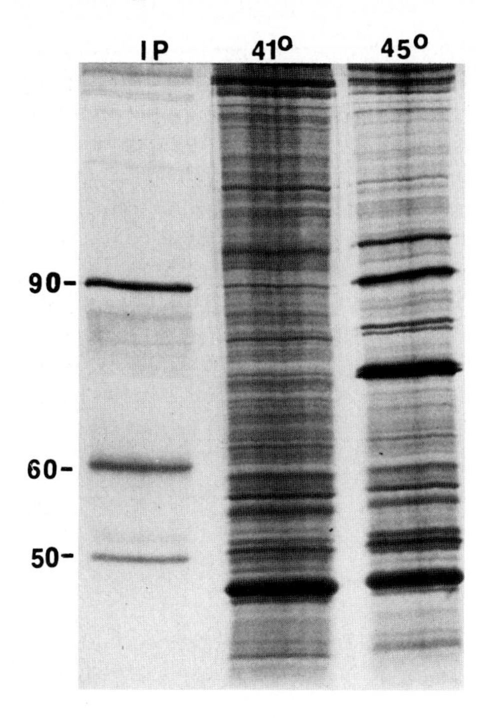

Fig. 3. Analysis of proteins labeled with ^{35}S-methionine under conditions of heat shock. Cells were incubated at 41° C (*middle lane*) or 45° C (*right lane*) for 3 h and incubated with ^{35}S-methionine for 2 h at each respective temperature. Total cellular proteins were electrophoresed on 7.5% polyacrylamide gels. The *left lane* shows an ^{35}S-methionine immunoprecipitate which contains pp50, pp90, and pp60^{src}

cytoplasmic protein after separation of cell fractions by homogenization in hypotonic solutions or extraction with detergent-containing cytoskeletal stabilizing buffers (YONEMOTO et al. 1982). Using monoclonal antibodies prepared against this protein, we have not observed any specific localization within the cytoplasm. Only a small fraction of pp90 is associated with pp60^{src}. pp90 is a phosphoprotein which contains several phosphorylated tryptic peptides, all of which contain phosphoserine (BRUGGE et al. 1981). Neither the synthesis nor the phosphorylation of pp90 is detectably altered by RSV-induced transformation.

OPPERMANN et al. (1981 b) contributed the interesting finding that pp90 is identical to one of several proteins whose synthesis increases after incubation of cells at elevated temperatures (heat shock) or other "stress-inducing" conditions (see Fig. 3). The heat shock proteins are highly conserved proteins which can also be found in lower eukaryotes and bacteria (SCHLESINGER et al. 1982). The genes encoding these proteins have been cloned from *Drosophila* and some other species; however, the functional role of these proteins in the heat shock response is not known. There are no obvious overlaps between the cellular events which occur during the heat shock response and virus-induced transformation which would provide clues to the function of pp90 in either system. We have not detected any change in the binding of pp90 to pp60^{src} when RSV-induced cells are incubated under heat shock conditions (YONEMOTO and BRUGGE, unpublished results).

The synthesis of pp90 is also sensitive to glucose starvation, exhibiting a rapid reduction in synthesis following deprivation of glucose (KASAMBALIDES and LANKS 1979; LANKS et al. 1982). Recently, pp90 has been shown to associate with another group of cellular proteins. In collaboration with Dr. David Toft, we have shown that the 90K protein found in association with the cytoplasmic, nonactivated form of the progesterone receptor (DOUGHERTY et al. 1984; SULLIVAN et al. 1985) is indistinguishable from the pp90 protein which complexes with pp60$^{v\text{-}src}$ (SCHUH et al., in press). W.J. WELCH and E.E. BAULIEU have also found that the 90K heat shock protein is indistinguishable from the progesterone receptor-binding protein (unpublished results). No functional role has been assigned to pp90 in the processing of the glucocorticoid receptor complex.

The purification of pp90 proved to be relatively straightforward, since large amounts of pp90 are present in uninfected cells and its synthesis can be induced several fold by various stress conditions (WELSH and FERAMISCO 1982). With this purified protein, several groups have prepared rabbit antiserum to pp90 (KELLEY and SCHLESINGER 1978; BRUGGE et al. 1983; WELSH and FERAMISCO, personal communication). We have also prepared a monoclonal antibody to pp90. When this antibody was used to immunoprecipitate pp90 from uninfected cells, pp50 was found to co-immunoprecipitate with pp90 (BRUGGE et al. 1982). This form of pp50 does not contain the phosphotyrosine-specific peptides present in molecules of pp50 bound to pp60^{src} in RSV-transformed cells. This suggests that a small percentage of pp90 molecules are bound to pp50 in the absence of pp60^{src}.

6.2 pp50

pp50 is not an abundant cellular protein and can only be detected in uninfected cells using two-dimensional protein analysis of ^{32}P-labeled proteins. A single species of phosphorylated pp50 is detectable in uninfected cells. This protein species contains phosphoserine. After transformation by RSV, an additional species of pp50 is detectable which migrates with a mobility consistent with the addition of one negative-charge unit (BRUGGE and DARROW 1982; GILMORE et al. 1982). Phosphopeptide and phospho-amino acid analyses of these pp50 species indicated that the transformation specific species of pp50 contains both phosphoserine and phosphotyrosine. The pp50 species which is precipitated with antibody to pp60^{src} comigrates with the transformation-specific species of pp50 and contains both phosphotyrosine and phosphoserine (BRUGGE and DARROW 1982). Three possible explanations for the appearance of the phosphotyrosine-containing species of pp50 after RSV-induced transformation are the following: (a) The phosphotyrosine-containing species of pp50 is short-lived in uninfected cells; binding of this pp50 species to pp60^{src} stabilizes it within the cell; (b) RSV infection activates a protein kinase which phosphorylates pp50 on tyrosine; or (c) pp60^{src} directly phosphorylates pp50.

Our experimental evidence supports possibility (c). First, there is the circumstantial association of pp60^{src} in a complex with pp50. Second, analysis of the phosphorylation of pp50 in cells infected with mutant viruses containing temperature-sensitive defects in the *src* gene (ts^{src}) suggests that pp50 is not phosphorylated by a protein kinase which is activated by pp60^{src}. When chicken cells are infected at 41° C with ts^{src} mutant viruses and maintained at 41° C such that pp60^{src} is never activated, pp50 is still phosphorylated on tyrosine (Brugge, unpublished results). If pp60^{src} activates another protein kinase which phosphorylates pp50, no such phosphorylation would be expected under these conditions. This observation that pp50 contains phosphotyrosine at 41° C in ts^{src} mutant virus-infected cells would also tend to rule out possibility (c), that pp60^{src} directly phosphorylates pp50; however, since the pp50:pp60^{src}:pp90 complex is very stable in ts^{src} mutant-virus infected cells at the nonpermissive temperatures, phosphorylation of pp50 which is stably associated with pp60^{src} might take place under conditions in which other substrates which are not physically complexed with pp60^{src} are not phosphorylated. Indeed, the stable binding of pp90 and pp50 could be responsible for the reduced phosphorylation of the other substrates seen in ts virus-infected cells at the nonpermissive temperature. While this latter evidence is not straightforward, taken together, the experimental findings indicate that pp50 might be directly phosphorylated by pp60^{src}. Further support for this possibility awaits the in vitro demonstration that pp50 can be phosphorylated by pp60^{src} on the same peptide which is phosphorylated in vivo.

Fractionation of pp50 into particulate and soluble fractions by homogenization of cells preincubated in hypertonic buffers indicated that pp50 is a highly soluble cellular protein. Like pp90, it also possesses a low isoelectric point of approximately 5.4 (Brugge and Darrow 1982; Gilmore et al. 1982). As pointed out above, pp50 is bound to pp90 in uninfected cells. The proportion of pp50 bound to pp90 is not known; however, the phosphorylated form of pp50 from uninfected cells was found to copurify with pp90 through several chromatographic steps, suggesting that most of the phosphorylated pp50 may be bound to pp90. In contrast, only a very small proportion of the total cellular pp90 is associated with pp50.

Analysis of partial proteolytic peptides generated with V8 protease, elastase, or chymotrypsin, indicated that the pp50 protein is not as highly conserved as the pp90 protein (Oppermann et al. 1981 b).

There are several lines of evidence which indicates that the tyrosine phosphorylation of pp50 does not play a crucial role in either the interaction of pp90 and pp50 with pp60^{src} or cellular transformation.

1. pp50 is not phosphorylated on tyrosine in mouse or rat cells transformed by pp60^{src} (Oppermann et al. 1981 b; Gates and Brugge, unpublished results).
2. pp50 is phosphorylated at the nonpermissive temperature in cells infected with viruses which contain ts defects within the *src* gene (Brugge and Darrow 1982).
3. pp50 is not phosphorylated at either 35° or 41° C in cells infected with a virus (CH119, Bryant and Parsons 1982) derived by deletion of *src* sequences from amino acids 173–227 (Lipsich and Brugge, unpublished results). This virus displays a ts-transformed phenotype. Thus, pp50 is not phosphorylated in cells which display transformed phenotype.

Taken together, these results suggest that phosphorylation of pp50 is not essential for either transformation or the association and dissociation of pp60^{src}

with pp90 and pp50. Therefore, it does not appear that phosphorylation of pp50 is the primary role of the interaction between these three proteins. It seems more likely that the avian pp50 molecule contains a tyrosine residue which is accessible to the catalytic site of pp60^{src} during complex formation.

7 Sites of pp60^{src} Which Interact with pp90 and pp50

Identification of the sites of pp60^{src} which interact with pp50 and pp90 may be useful for understanding the functional role of this complex. Several indirect and direct approaches suggest that pp50 and pp90 interact with the carboxyl half of pp60^{src}.

7.1 Analyses of Viruses Carrying Mutations Within the src Gene

Table 1 summarizes the properties of many mutant viruses which have been investigated with respect to the interaction of pp60^{src} with pp50 and pp90.

All of the mutant viruses which were selected in vivo for ts expression of focus formation displayed elevated levels of the complex at the nonpermissive temperature. While the lesions responsible for the mutant phenotype of most of these viruses have not been precisely mapped, the analysis of recombination with deletion mutants has indicated that the mutations present within these viruses map within the COOH half of the *src* gene (FINCHAM et al. 1982).

The analysis of viruses carrying deletion mutants within the *src* gene indicate that no sequences within the amino half of the pp60^{src} molecule are required for binding to pp50 and pp90. pp60^{src} molecules from mutant 312, which is deleted from amino acids 15–254, and mutant 314, deleted from amino acids 2–81, bind to pp50 and pp90. Although sequences within the amino half of pp60^{src} are not essential for binding, these sequences appear to influence the dissociation of this complex, since the majority of pp60^{src} is found to be stably associated with pp90 and pp50 in cells infected with mutant 312 (deleted from amino acids 15–264) and mutant 18-3 (deleted from 169–264) (CROSS et al. 1984; E. GARBER, F. CROSS, H. HANAFUSA, unpublished results).

The only mutations which appear to prevent complex formation are three mutants which contain frame shifts at the carboxyl terminus of pp60^{src} (CHis 1545, CHdl 300, CHis 1511; T. PARSONS et al. 1984 and unpublished material). No complex has been detected in cells infected with these viruses. This evidence further supports the importance of the COOH half of pp60^{src} in the binding to pp50 and pp90.

7.2 Analyses of the Complex Using Antibodies to Specific Regions of pp60^{src}

Since binding of pp50 and pp90 to pp60^{src} may block access of antibody molecules to the regions of pp60^{src} which are bound to pp50 and pp90, the use of antibody molecules of defined specificity can also be used to map the sites

Table 1. Summary of the properties of mutant viruses

	Amino acid alteration	90 and 50 bound	Morph	(IgG) Kinase	Membrane association	Myris-tylation
wt SR-RSV	–	1%–5%	Round	1.0	+	+
ts 68	?	>90% 41° C	Ts	Ts	Ts	+
tsLA90	?	>90% 41° C	Ts	Ts	ND	ND
307	Del 15–27 Insert PQIW	1%–5%	Round	1.0	+	+
309	Del 15–81 Insert PDL	1%–5%	Fusiform	0.8	+	+
314	Del 2–81 Insert DL	1%–5%	Flat	0.6	–	–
315	Del 2–15 Insert DLG	1%–5%	Flat	1.0	–	–
300	Del 2–4-sub Insert NRSG	1%–5%	Flat	1.0	–	–
312	Del 15–264	>90%	Flat	0.04	–	+
18-3	Del 169–264 Insert PQICG	>90%	Flat	0.08	–	+
CHis1545	Frameshift At 516	<0.1%	Flat	–	Labile	ND
CH 1	433, thr→ala	0.1%–1%	Flat	–	ND	ND
CH119	Del 173–227	1%–5%,	Ts	1.0 35° C 0.5 41° C	ND	ND

Del, deletion; ND, not determined; SR, Schmidt-Ruppin strain
References: Ts 68 (KAWAI and HANAFUSA 1971); 300, 307, 309, 312, 314, 315, 18-3, (CROSS et al. 1984; E. GARBER, D. PELLMAN, F. CROSS and H. HANAFUSA, unpublished results); CHis1545 (PARSONS et al. 1984) CH1 (BRYANT and PARSONS 1983) CH119 (BRYANT and PARSONS 1982)

of interaction between pp50, pp60^{src}, and pp90. SEFTON and WALTER (1982) have shown that antibody prepared against the carboxyl terminal six amino acids of pp60^{src} precipitate the complexed form of pp60^{src} inefficiently. This suggests that pp50 and pp90 may partially mask the recognition of this region of pp60^{src}. In addition, TAMURA et al. (1983) have selected specific antibody populations from TBR serum by adsorption to affinity columns containing synthetic peptides from various regions of the *src* protein. With the exception of a peptide which spanned amino acids 155–160, the antibodies which did not precipitate complex-bound pp60^{src} were from the carboxyl half of pp60^{src}.

8 Model for the Interaction Between pp50 and pp90

Figure 4 presents a model for describing the events following synthesis of pp60^{src}. This model is based on the evidence which is described in the preceding sections on the interaction between pp50, pp60^{src}, and pp90.

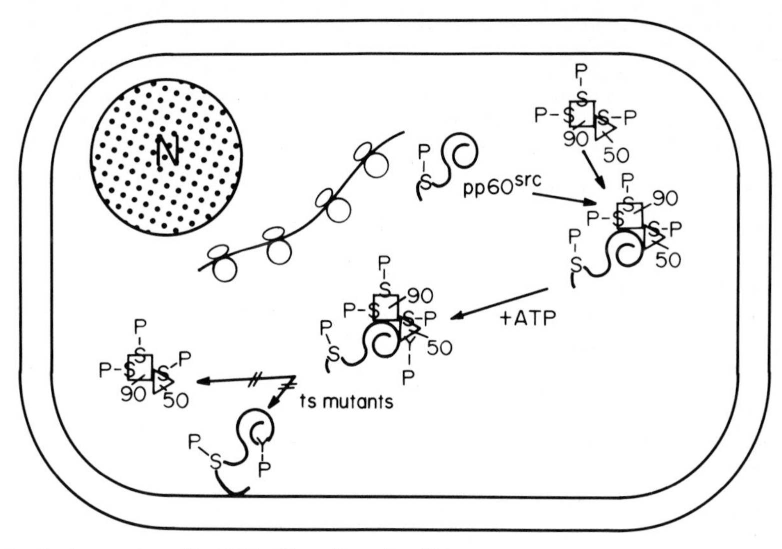

Fig. 4. Model for the interaction of pp60^{src} with pp90 and pp50

1. Newly synthesized molecules of pp60^{src} which are phosphorylated on serine bind to pp50 and pp90. pp50 and pp90 bind to the catalytic domain of phosphotransferase activity contained within the carboxyl half of pp60^{src}. pp50 and pp90 may enter the complex together, since these two proteins are associated in a protein complex in uninfected cells. Molecules of pp50 which enter this complex are not phosphorylated on tyrosine, but contain phosphoserine. The binding of pp50 and pp90 with pp60^{src} does not take place in association with any membrane or cytoskeletal cellular components.
2. Phosphorylation of pp50 on tyrosine occurs after formation of the trimolecular complex.
3. Dissociation of the pp50:pp60^{src}:pp90 complex and of pp60^{src} with the plasma membrane. At some time before of shortly after the association of pp60^{src} with the plasma membrane, pp60^{src} is phosphorylated on tyrosine residue 416. The precise order of events which promote complex dissociation and membrane binding are unknown.

Step 3 is blocked in mutants containing ts defects in the *src* gene. We have examined more than ten independent isolates of mutant viruses which induce a temperature-dependent transformed phenotype. Cells infected with all of these viruses show elevated levels of complex binding at the nonpermissive temperature, suggesting that complex dissociation is prevented or slow in these virus-infected cells.

9 Possible Functional Roles of the Interaction Between pp50, pp60^{src}, and pp90

The temporal sequence of events which occurs following the synthesis of pp60^{src} suggests that pp90 and pp50 may be involved in processing events which take place before pp60^{src} reaches the plasma membrane. The functional nature of these events is not clear, however; several possible functions are discussed below. While experimental evidence does not overwhelmingly support any one possibility, none of them be totally disregarded.

9.1 Transport of pp60^{src} to the Plasma Membrane

The evidence that pp50 and pp90 bind to newly synthesized pp60^{src} before it reaches the plasma membrane suggests that pp50 and pp90 may direct pp60^{src} to the plasma membrane. Unlike many plasma proteins, pp60^{src} is not translated on membrane-bound polyribosomes (LEE et al. 1979) or transported to the membrane via the endoplasmic reticulum and Golgi apparatus. pp50 and pp90 could provide a transport or docking system for pp60^{src} and the other tyrosine-kinase transforming proteins which are also associated with the plasma membrane.

The mechanism whereby pp50 and pp90 might facilitate pp60^{src} transport would not appear to involve an association with the cytoskeletal apparatus since the pp50:pp60^{src}:pp90 complex is soluble in buffers which stabilize the cytoskeleton. pp50 and pp90 are not membrane proteins and thus would not appear to have a stable interaction with the plasma membrane.

Several mutant forms of pp60^{src} which have an altered interaction with the plasma membrane have been isolated. Many of the mutant pp60^{src} proteins which do not stably associate with the plasma membrane are bound in a stable complex with pp90 and pp50. The absence of membrane association may be due to defects in the dissociation of pp90 and pp50 from pp60^{src}. One could speculate that this transport system has evolved to be highly sensitive to the configuration of the proteins which it transports. As a consequence, mutant proteins might form a nondissociable complex which cannot be transferred to the usual pp60^{src} localization in the cell.

Although evidence from these ts mutant viruses provides some support for the possibility that pp90 and pp50 might be involved in transit to the membrane, it does not appear that stable association with the membrane is required for complex dissociation since the *src* proteins from mutants which are not stably associated with the membrane appear to bind to, and dissociate from, pp50 and pp90 with similar kinetics as with pp60^{src} (mutants 300, 314, and 315; CROSS et al. 1984; E. GARBER, F. CROSS, D. PELLMAN, H. HANAFUSA, unpublished results).

Mutant forms of pp60^{src} which do not bind to pp50 and pp90 should be very useful for examining the function of the pp90, pp50, pp60^{src} complex. The only mutant viruses which show this phenotype were derived by linker insertions at the 3′ end of the *src* gene (PARSONS et al. 1984). Preliminary biochemical analysis of the localization of the *src* protein from one mutant, CHis 1545, suggests that pp60^{src} does not establish a stable association with the plasma membrane or any cellular membranes in cells infected with this mutant virus (LIPSICH and BRUGGE, unpublished results). Immunofluorescence studies of cells infected with CHis 1545 revealed an asymetric perinuclear pattern of fluorescence with no apparent pp60^{src} localization at the plasma membrane, adhesion plaques or cell:cell junctions (T. PARSONS, unpublished material). This evidence is consistent with the possibility that pp60^{src} does not reach the plasma membrane in the absence of pp90 and pp50 binding; however, since this mutant form of the *src* protein is defective in kinase activity and other pp60^{src} mediated functions, one cannot conclude that pp90 and pp50 binding, per se, is responsible for transport of pp60^{src} to the plasma membrane.

Another possible role for pp50 and pp90 which is similar, but not identical to the transport function, is to maintain the solubility of pp60^{src} within the soluble cellular material. This does not constitute transport, but a means to preserve the solubility of the *src* before its interaction with the plasma membrane. In this case, the association of pp50 and pp90 with newly synthesized molecules of pp60^{src} would facilitate transport of pp60^{src} to the membrane by maintaining the solubility of pp60^{src}; however, "directed" transport per se would not be the primary function of this complex. According to this scenario, one might predict that all molecules of pp60^{src} which are not associated with the plasma membrane, whether newly synthesized or otherwise, would be bound to pp90 and pp50. The fact that the soluble form of pp60^{src} from cells infected with ts mutants of RSV is entirely associated with pp90 and pp50 at the non-permissive temperature supports this possibility. However, the following, more compelling, lines of evidence do not support this proposed function:

1. When the soluble fraction of cellular proteins from cells infected with nondefective RSV were sedimented on glycerol gradients, approximately half of the soluble pp60^{src} molecules were not complexed with pp90 and pp50. This suggests that pp90 and pp50 are not essential for maintaining the solubility of pp60^{src} in this fraction, although it is possible that the pp60^{src} molecules which were not "complexed" could have been released from the plasma membrane during the homogenization procedure (BRUGGE and DARROW, unpublished results).
2. The *src* proteins from several mutant viruses constructed by CROSS and coworkers which fractionate as soluble, cytoplasmic proteins with no apparent association with the plasma membrane, are not stably complexed with pp90 and pp50 (mutants 300, 314, 315; CROSS et al. 1984; E. GARBER, F. CROSS, D. PELLMAN, H. HANAFUSA, unpublished results).

9.2 Attachment of Myristate to pp60^{src}

Another possible function of the interaction between pp50, pp60^{src}, and pp90 is catalysis of the linkage of myristate to newly synthesized molecules of pp60^{src}. It was recently shown that the 14-carbon fatty acid, myristate, is covalently bound to pp60^{src} (SEFTON et al. 1982; CROSS et al. 1984; B. SEFTON, unpublished results; A. SCHULTZ, S. OROSZLAN, E. GARBER, H. HANAFUSA, unpublished results). It has been speculated that this fatty-acid residue might facilitate the association of pp60^{src} with the plasma membrane. That this is the functional role of myristate attachment is not clear, since proteins like the catalytic subunit of cAMP-dependent protein kinase, which also contain myristate, do not have a detectable interaction with the plasma membrane (CARR et al. 1982).

Experimental evidence does not provide strong support for the possibility that pp90 and pp50 are responsible for fatty acylation of pp60^{src}. (a) Mutant forms of pp60^{src} from cells infected with viruses (Table 1) which do not contain myristate associate with and dissociate from pp90 and pp50 in the same pattern as nondefective pp60^{src} (CROSS et al. 1984; E. GARBER, F. CROSS, H. HANAFUSA, unpublished results; see mutants 300, 314, and 315). (b) The mutant form of pp60^{src} from cells infected with mutant viruses NY68 and 18-3 is stably associated with pp90 and pp50 and contains myristate (CROSS et al. 1984; and E. GARBER, F. CROSS, and H. HANAFUSA, unpublished results). This evidence, taken together with point (a) would suggest that the addition of myristate to pp60^{src} is not responsible for dissociation of pp50 and pp90 from pp60^{src}.

9.3 Phosphorylation of pp50

The primary function of the interaction between pp50, pp60^{src}, and pp90 could be pp60^{src}-mediated phosphorylation of pp50. One could speculate that this phosphorylation of pp50 might be responsible for eliciting one of the many pleiotropic changes which occur after viral transformation. Although most enzyme-substrate complexes are not stable enough to be detected by immunoprecipitation, complex formation in RSV-transformed cells could reflect variations in the kinetics of this enzyme-substrate interaction, perhaps due to the influence of pp90 or to changes in the amino acid sequence in the active site of viral pp60^{src} compared with cellular pp60^{src}. We have observed that polymorphism in the *src* genes from different strains affects the half-life of the complex in RSV-transformed cells (BRUGGE et al. 1983).

Although the evidence which we have obtained supports the possibility that pp50 is phosphorylated by pp60^{src} (see above), there are several unique features of this enzyme substrate interaction which suggest that phosphorylation of pp50 may not be the function of this complex.

1. pp60^{src} does not form a stable, immunoprecipitable complex with any of the other candidate substrates of pp60^{src}-mediated phosphorylation, such as vinculin, (SEFTON et al. 1981) or the major 34K–36K substrate (RADKE and MARTIN 1979; ERIKSON and ERIKSON 1980). If changes in enzyme kinetics are responsible for complex formation, one might expect this to affect the interaction with other substrates (unless pp90 is responsible for altering the kinetics).

2. Complex formation occurs in transformed mammalian cells, but pp50 is not phosphorylated during the interaction (OPPERMANN et al. 1981a). This result suggests that pp50 phosphorylation is not the function of the complex and is not required for transformation (unless the absence of detectable phosphorylation is a consequence of a difference in the turnover of phosphate on the tyrosine residue of pp50). The mammalian pp50 protein might have diverged such that this interaction does not involve or require phosphorylation in order to perform its primary function.

9.4 *Regulation of the Phosphotransferase Activity of pp60^{src}*

Complex formation could be involved in regulating the functional activity of pp60^{src}. Elaborate systems of regulation, both positive and negative, have been described for cAMP-dependent protein kinases (reviews, KREBS and BEAVO 1979; COHEN 1982). Several levels of control involving either phosphorylation, dephosphorylation, allosteric cofactors, or physical association with inhibitory proteins have been described. It is reasonable to suspect that the cellular tyrosine-specific protein kinases are also highly regulated. It is possible that pp90 and pp50 could play regulatory roles in the expression of pp60^{src} activity. There are several features of the interaction between the regulatory (R) and catalytic (C) subunits of the cAMP-dependent protein kinase (cAMP-PK) and heat-stable (hs) inhibitor protein that are noteworthy in comparison to the pp50, pp60^{src}, and pp90 interaction. While not entirely analogous, this system illustrates several concepts regarding the regulation of a cellular protein kinase.

1. The C subunit of cAMP-PK is inactive when bound to either its R subunit or the hs inhibitory protein (WALSH and KREBS 1973). Analogously, pp90 and pp50 appear to block pp60^{src}-mediated phosphorylation of other substrates.
2. Dissociation of the R subunits from one R_2C_2 complex activates the catalytic subunit (WALSH and KREBS 1983) as a protein kinase. Analogously, pp90 and pp50 could activate the enzymatic activity of newly synthesized pp60^{src}. The *src* protein produced in *E. coli* has detectable protein kinase activity, but the specific activity appears to be considerably lower than that of the protein produced in animal cells. pp50 and pp90 interaction could be important for proper processing of pp60^{src}.
3. There are two cellular R proteins (type I and type II) each of which have similar functions in the regulation of cAMP-PK activity; however, only the II-R protein is phosphorylated during the interaction (RANGEL-ALDAL and ROSEN 1976). The only difference between phosphorylated and unphosphorylated R proteins which has been detected is that the phosphorylated form of R has a slower rate of reassociation with C protein than the unphosphorylated R protein, and thus might have a small influence on the total cellular cAMP-PK activity. This might be analogous to pp50 in chicken vs mammalian cells. pp50 could have diverged such that it is not able to be phosphorylated by pp60^{src}, but still carries out the primary function of the interaction.
4. Mutant cell lines have been isolated which have no cAMP-PK activity. One of these lines has been shown to have a defect in the C protein which causes the formation of a nondissociable R_2C_2 complex (M. GOTTESMAN, unpublished results). This might be analogous to the ts^{src} mutants of RSV.

There is no direct evidence that pp90 and pp50 regulate the functional activity of pp60^{src} (except in a negative fashion when bound). No in vitro reconstruction system has been developed which would allow more direct investigations of the influence of pp90 and pp50 on the activity of pp60^{src}. We have also not yet detected activity after dissociation of pp90 and pp50 from pp60^{src} in vitro. It is possible that the conditions used for dissociation (0.5 m KSCN) resulted in loss of protein kinase activity although similar treatment of unbound pp60^{src} did not result in loss of activity. It is difficult to draw meaningful conclusions from this type of analysis except in cases in which a positive effect is observed. Therefore, the development of an in vitro system for the association and dissociation of pp60^{src}, pp50, and pp90 could be useful for analysis of the functional role of the pp50:pp90 interaction with pp60^{src}. Equally useful would be a mutant virus which does not interact with pp90 and pp50, but is not totally defective for kinase or transformation activity so that one could sort out the specific contribution of pp90 and pp50 to the processing and activity of pp60^{src}.

10 Interaction of pp90 and pp50 with Nonviral Proteins

All of the viral transforming genes which encode tyrosine-specific protein kinases are homologous to highly conserved cellular genes. These cellular genes have been detected in chromosomal DNA from lower eukaryotic species (such as *Drosophila*) as well as from humans. The protein products of these genes are expressed in normal cells and are similar antigenically, structurally, and functionally to their viral counterparts (review, BISHOP and VARMUS 1982). The presence of these proteins in normal cells raises the question of whether the cellular homologs of the viral tyrosine-kinase transforming proteins also form a transient complex with pp50 and pp90. This question has been difficult to address because of the low levels of the viral homologs in uninfected cells. For instance, the cellular *src* protein (pp60^{src}) is present at 30- to 40-fold lower levels in uninfected chicken embryo fibroblasts (CEF) than pp60$^{v\text{-}src}$ in RSV-transformed cells (COLLETT et al. 1978). If only 1%–5% of this protein is present in a complex with pp90 and pp50 (as observed with pp60$^{v\text{-}src}$, in viral transformed cells), then it would be necessary to detect a protein whose level was approximately 0.00015% of total cellular protein in order to detect pp90 and pp50 association with c-*src* protein. Thus, the inability to detect pp90 and pp50 in pp60$^{c\text{-}src}$ immunoprecipitates did not rule out the possibility of an interaction between these proteins. Recently, IBA et al. (1984) have constructed a virus containing the c-*src* gene in the position of v-*src* gene within RSV. Infection of cells with this virus leads to synthesis of high levels of the c-*src* protein in the absence of transformation. These investigators have not detected pp90 and pp50 in pp60$^{c\text{-}src}$ immunoprecipitates from these cells. JOHNSON et al.(1985) have expressed the c-*src* gene on a plasmid vector containing the murine leukemia virus LTR which allows high levels of c-*src* expression in mouse cells. Using a stable line of cells transfected with this plasmid which expresses approximately 20-fold higher levels of c-*src* than found in normal 3T3 cells, we have detected low levels of pp90 present in immunoprecipitates using monoclonal antibodies to pp60$^{c\text{-}src}$ (SCHUH and BRUGGE, unpublished results). The presence of a 50K cleavage product of pp60^{src} has prevented identification of pp50 in these immunoprecipitates. In these cells, as well as in chicken cells infected with the c-*src*-containing virus, the ratio of pp60$^{c\text{-}src}$/pp90 is much higher than the ratio of pp60$^{v\text{-}src}$/pp90 in viral transformed cells. We are presently addressing the question of whether the difference in the levels of the pp60$^{c\text{-}src}$:pp50:pp90 complex is due to an increase in the turnover of this complex, an increased half-life of pp60$^{c\text{-}src}$ in these cells expressing high levels of c-*src*, a smaller population of the c-*src* protein which interacts with pp50 and pp90, or a less stable interaction between the c-*src* protein and pp90 and pp50. It is also possible that the small number of complexed pp60$^{c\text{-}src}$ proteins represent mutant forms of the c-*src* protein which have been generated in these high-expression systems. Iba et al. (1984) have found a high frequency of transforming viruses which are generated from their nontransforming, c-*src*-containing viruses.

It seems unlikely that the viral-encoded forms of the *src*, *yes*, *fps*, *fes*, and *fos* gene products would all interact with pp50 and pp90 and that none of the cellular homologs would bind to these proteins. If this were the case, then one must postulate a convergent mutation in all of these genes which caused these proteins to interact with pp50 and pp90. It seems more likely that amino acid differences between the cellular and viral forms of these proteins could alter some aspect of the processing of this complex which might facilitate its detection in virus-transformed cells (i.e., to alter the kinetics of the interaction with pp50 and pp90 such that the complex has a longer half-life in the viral-transformed cells).

11 Future Directions

It is clear that the time and place of the interaction between pp60^{src}, pp50, and pp90 has been defined. However, the function of this interaction remains unknown. Circumstantial evidence strongly supports a transport or membrane-directing function for this complex; however, without direct demonstration of this role, this model remains speculation. Two areas of investigation may lead to a better definition of the events involved in association and dissociation of this complex. The most useful would be an in vitro system for reconstruction of the association and dissociation of pp60^{src} analogous to the enlightening work on the processing of proteins which are translated in association with membrane-bound polysomes (Blobel 1982). The complexities in the development of such a system are obvious, yet the rewards would be great.

Further elucidation of the function of pp50:pp60^{src}:pp90 complex might also come from the generation and analysis of mutant *src* proteins which have a defective interaction with pp90 and pp50. The previously characterized *src* proteins from ts mutant viruses which appear to bind irreversibly with pp90 and pp50 are not particularly useful for functional studies. However, the construction of mutations, like CHis 1545, which prevent binding of pp90 and pp50 will help to elucidate the role of this complex in the processing of pp60$^{v\text{-}src}$ and perhaps pp60$^{c\text{-}src}$ and other cellular plasma membrane-associated tyrosine kinases which are not translated on membrane-bound polysomes. It is likely that pp90 and pp50 have a common function in the processing of this class of cellular and viral proteins and elucidation of the function of this interaction may reveal a new, unidentified cellular process.

Acknowledgments. I am grateful to all colleagues who contributed unpublished results and for discussions of the complex. I also thank SUSAN SCHUH and LEAH LIPSICH for critical reading of the manuscript and PHYLLIS LEDER and KATHY DONNELLY for excellent secretarial assistance.

References

Adkins B, Hunter T, Sefton BM (1982) The transforming proteins of PRCII virus and Rous sarcoma virus form a complex with the same two cellular phosphoproteins. J Virol 43:448–455

Bishop JM, Varmus H (1982) Functions and origins of retroviral transforming genes. In: Weiss R, Teich N, Varmus H, Coffin J (eds) RNA tumor viruses. Cold Spring Harbor Press, New York, pp 999–1108

Blobel G (1982) Regulation of intracellular protein traffic. Cold Spring Harbor Symposia on quantitative biology, vol XLVI. Cold Spring Harbor Press, New York, pp 7–16

Brugge JS, Darrow D (1982) Rous sarcoma virus-induced phosphorylation of a 50,000 molecular weight cellular protein. Nature 295:250–253

Brugge JS, Erikson RL (1977) Identification of a transformation-specific antigen induced by an avian sarcoma virus. Nature 269:346–348

Brugge J, Erikson E, Erikson RL (1981) The specific interaction of the Rous sarcoma virus transforming protein, pp60^{src}, with two cellular proteins. Cell 25:363–372

Brugge JS, Darrow D, Lipsich LA, Yonemoto Y (1982) The association of the transforming protein of Rous sarcoma virus with two cellular phosphoproteins. In: Rauuscher R (ed) Oncogenes: evaluation of basic findings and clinical potential. Liss, New York, pp 135–148

Brugge JS, Yonemoto W, Darrow D (1983) Interaction between the Rous sarcoma virus transforming protein and two cellular phosphoproteins: Analysis of the turnover and distribution of this complex. Mol Cell Biol 3:9–19

Bryant D, Parson JT (1982) Site directed mutagenesis of the *src* gene of Rous sarcoma virus: construction and characterization of a deletion mutant temperature sensitive for transformation. J Virol 44:683–691

Bryant D, Parson JT (1983) Site-directed point mutation in the src gene of Rous sarcoma virus results in an inactive src gene product. J Virology 45:1211–1216

Carr SA, Biemann K, Shoji S, Parmelee DC, Titani K (1982) N-teradecanoyl is the NH2-terminal blocking group of the catalytic subunit of cyclic AMP-dependent protein kinase from bovine cardiac muscle. Proc Natl Acad Sci USA 79:6128–6131

Cohen P (1982) The role of protein phosphorylation in neural and hormonal control of cellular activity. Nature 296:613–620

Collett MS, Erikson RL (1978) Protein kinase activity associated with the avian sarcoma virus *src* gene product. Proc Natl Acad Sci USA 75:2021–2024

Collett MS, Brugge JS, Erikson RL (1978) Characterization of a normal avian cell protein related to the avian sarcoma virus transforming gene product. Cell 15:1363

Courtneidge SA, Bishop JM (1982) Transit of pp60$^{v\text{-}src}$ to the plasma membrane. Proc Natl Acad Sci USA 79:7117–7121

Courtneidge SA, Levinson AD, Bishop JM (1980) The protein encoded by the transforming gene of avian sarcoma virus (pp60^{src}) and homologous protein in normal cells (pp60$^{proto\text{-}src}$) are associated with the plasma membrane. Proc Natl Acad Sci USA 77:3783–3787

Cross FR, Hanafusa H (1983) Local mutagenesis of Rous sarcoma virus: the major sites of tyrosine and serine phosphorylation are dispensable for transformation. Cell 34:597–608

Cross F, Garber E, Pellman D, Hanafusa H (1984) A N-terminal region of p60^{src} is required for its myristylation and membrane association, and for cell transformation. Mol Cell Biol 4:1834–1842

Dougherty JJ, Puri RK, Toft DO (1984) Polypeptide components of two 8S forms chicken oviduct progesterone receptor. J Biol Chem 259:8004–8009

Erikson E, Erikson RL (1980) Identification of a cellular protein substrate phosphorylated by the avian sarcoma virus-transforming gene product. Cell 21:829–836

Fincham V, Chiswell DJ, Wyke J (1982) Mapping of nonconditional and conditional mutants in the *src* gene of Prague strain Rous sarcoma virus. Virology 116:72–83

Gilmore T, Radke K, Martin GS (1982) Tyrosine phosphorylation of a 50K cellular polypeptide associated with the Rous sarcoma virus-transforming protein, pp60^{src}. Mol Cell Biol 2:199–206

Hanafusa H (1977) Cell transformation by RNA tumor viruses. In: Fraenkel-Conrat H, Wagner RP (eds) Comprehensive virology. Plenum, New York, pp 401–483

Hunter T, Sefton B (1980) Transforming gene product of Rous sarcoma virus phosphorylates tyrosine. Proc Natl Acad Sci USA 77:1311–1315

Iba H, Takeya T, Cross F, Hanafusa T, Hanafusa H (1984) Rous sarcoma virus variants which carry the cellular *src* gene instead of the viral *src* gene cannot transform chicken embryo fibroblasts. Proc Natl Acad Sci USA 81:4424–4428

Johnson PJ, Coussens PM, Danko AV, Shalloway D (1985) Overexpressed pp60$^{c\text{-}src}$ can induce focus formation without complete transformation of NIH 3T3 cells. Mol Cell Biol 4:454–467

Kasambalides EJ, Lanks KW (1979) Patterns of proteins synthesized by nonproliferating murine L cells. Exp Cell Res 118:269–275

Kawai S, Hanafusa H (1971) The effects of reciprocal changes in the temperature on the transformed state of cells infected with a Rous sarcoma virus mutant. Virology 46:470–479

Kelley PM, Schlesinger MJ (1978) The effect of amino acid analogues and heat shock on gene expression in chicken embryo fibroblasts. Cell 15:1277–1286

Krebs EG, Beavo JA (1979) Phosphorylation and dephosphorylation of enzymes. Ann Rev Biochem 48:923–959

Krueger JK, Garber EA, Chin SS-M, Hanafusa H, Goldberg AR (1984) Size variant pp60^{src} proteins of recovered avian sarcoma viruses interact with adhesion plaques as peripheral membrane proteins: effects on cell transformation. Mol Cell Biol 4:454–467

Lanks KW, Kasambalides EJ, Chinkers M, Brugge JS (1982) A major cytoplasmic glucose-regulated protein is associated with the Rous sarcoma virus pp60^{csrc} protein. J Biol Chem 257:8604–8607

Lee JS, Varmus HE, Bishop JM (1979) Virus-specific messenger RNAs in permissive cells infected by avian sarcoma virus. J Biol Chem 254:8015–8022

Levinson AD, Oppermann H, Levintow L, Varmus HE, Bishop JB (1978) Evidence that the transforming gene of avian sarcoma virus encodes a protein kinase associated with a phosphoprotein. Cell 15:561–572

Lipsich LA, Cutt J, Brugge JS (1982) Association of the transforming proteins of Rous, Fujinami and Y73 avian sarcoma viruses with the same two cellular proteins. Mol Cell Biol 2:875–880

Lipsich LA, Lewis AJ, Brugge JS (1983) Isolation of monoclonal antibodies which recognize the transforming proteins of avian sarcoma viruses. J Virol 48:352–360

Mathey-Prevot B, Shibuya M, Samarut J, Hanafusa H (1984) Revertants and partial transformants of rat fibroblasts infected with Fujinami sarcoma virus. J Virol 50:325–334

Oppermann H, Levinson W, Bishop JM (1981 a) A cellular protein that associates with a transforming protein of Rous sarcoma virus is also a heat-shock protein. Proc Natl Acad Sci USA 78:1067–1071

Oppermann H, Levinson AD, Levintow L, Varmus HE, Bishop JM, Kawai S (1981 b) Two cellular proteins that immunoprecipitate with the transforming protein of Rous sarcoma virus. Virology 113:736–751

Parsons JD, Bryant D, Wilkerson V, Gilmartin G (1984) Site directed mutagenesis of Rous sarcoma virus: Identification of functional domains required for transformation. In: Van de Woude GF, Levine AJ, Topp WC, Watson JD (eds) Cancer cells 2: oncogenes and viral genes. Cold Spring Harbor, New York, pp 37–42

Purchio AF, Erikson E, Brugge JS, Erikson RL (1978) Identification of a polypeptide encoded by the avian sarcoma virus *src* gene. Proc Natl Acad Sci USA 75:1567–1571

Radke K, Martin GS (1979) Transformation by Rous sarcoma virus: effects of *src* gene expression on the synthesis and phosphorylation of cellular polypeptides. Proc Natl Acad Sci USA 76:5213–5216

Rangel-Aldao R, Rosen OM (1976) Mechanism of self-phosphorylation of adenosine 3′:5″-monophosphate-dependent protein kinase from bovine cardiac muscle. J Biol Chem 251:7526–7540

Schlesinger MJ, Ashburner M, Tissieres A (eds) (1982) Heat shock: from bacteria to man. Cold Spring Harbor Laboratory, New York

Schuh S, Yonemoto W, Brugge J, Bauer VJ, Riehl R, Sullivan WP, Toft DO. A 90,000 dalton binding protein common to both steroid receptors and the Rous sarcoma virus transforming protein pp60$^{v\text{-}src}$. J Biol Chem 260, in press

Sefton BM, Hunter T (1984) Tyrosine protein kinases. In: Greengard P, Robison GA (eds) Advances in cyclic nucleotide and protein phosphorylation research. Raven Press, New York, pp 195–226

Sefton BM, Walter G (1982) An antiserum specific for the carboxy terminus of the transforming protein of Rous sarcoma virus. J Virol 44:467–474

Sefton BM, Beemon K, Hunter T (1978) Comparison of the expression of the *src* gene of Rous sarcoma virus in vitro and in vivo. J Virol 28:957–971
Sefton BM, Hunter T, Ball EH, Singer SJ (1981) Vinculin: A cytoskeletal target of the transforming protein of Rous sarcoma virus. Cell 24:165–174
Sefton BM, Cooper JA, Trowbridge S, Scolnick EM (1982) The transforming proteins of Rous sarcoma virus, Harvey sarcoma virus and Abelson virus contain tightly bound lipid. Cell 31:465–474
Smart JE, Oppermann H, Czernilofsky AP, Purchio AF, Erikson RL, Bishop JM (1981) Characterization of sites for tyrosine phosphorylation in the transforming protein of Rous sarcoma virus ($pp60^{v\text{-}src}$) and its normal cellular homologue ($pp60^{c\text{-}src}$). Proc Natl Acad Sci USA 78:6013–6017
Sullivan WP, Vroman BT, Bauer VJ, Puri RK, Riehl RM, Pearson GR, Toft DO Isolation of a steroid receptor-binding protein from the chicken oviduct and production of monoclonal antibodies. Biochem (in press)
Tamura T, Bauer H, Birr C, Rudiger P (1983) Antibodies against synthetic peptides as a tool for functional analysis of the transforming protein $pp60^{src}$. Cell 34:587–596
Yonemoto W, Lipsich L, Darrow D, Brugge JS (1982) An analysis of the interaction of the Rous sarcoma virus transforming protein, $pp60^{src}$, with a major heat shock protein. In: Schlesinger MJ, Ashburner M, Tissieres A (eds) Heat shock proteins: from bacteria to man. Cold Spring Harbor Laboratory, New York, pp 289–298
Walsh DA, Krebs EG (1973) Protein kinases. In: Bayer P (ed) The enzymes, vol 8. Academic, New York, p 555
Welch WJ, Feramisco JR (1982) Purification of the major mammalian heat shock proteins. J Biol Chem 257:14949–14959

Microinjection Studies of Retroviral Polynucleotides

D.W. STACEY

1 Viral Structure 23
2 Microinjection of mRNA 24
3 Studies of Rare Biological Activities 26
4 *env* mRNA Packaging 27
5 DNA Injections 30
6 DNA Studies of Packaging 33
7 Protein Injections 34
8 Summary 36
References 36

The biological properties of retroviral macromolecules have been studied using the technique of microinjection together with genetic and molecular analyses. These studies not only provided insight into the major activities of viral RNA species, but also revealed rare events in which these molecules participate. The results provide a general correlation between viral genomic regions and biological activities of the virus. The general observations made with RNA injections are now being refined using recombinant DNA technology.

1 Viral Structure

The retroviral particle is composed of two full-genomic RNA molecules enclosed in a capsid and core particle made up primarily of products of the viral group-specific antigen (*gag*) gene. The lipid membrane surrounding the virion is derived from the host cell during viral budding and contains the viral envelope glycoprotein (*env*) gene product which is required for infectivity. Viral infection does not necessarily harm the cell and perhaps 1% of the mRNA of an infected cell is viral specific (BISHOP 1978; COFFIN 1979). Viral RNA within the cell is in the form of full genomic molecules along with smaller spliced species (HAYWARD 1977; WEISS et al. 1977; Fig. 1). Virion RNA, however, is primarily comprised of two 35S–40S full genomic molecules associated in a 50S–70S complex. The genomic molecule serves as template for reverse transcriptase, the

Department of Cell Biology, Roche Institute of Molecular Biology, Roche Research Center, Nutley, NJ 07110, USA

Current Topics in Microbiology and Immunology, Vol. 123

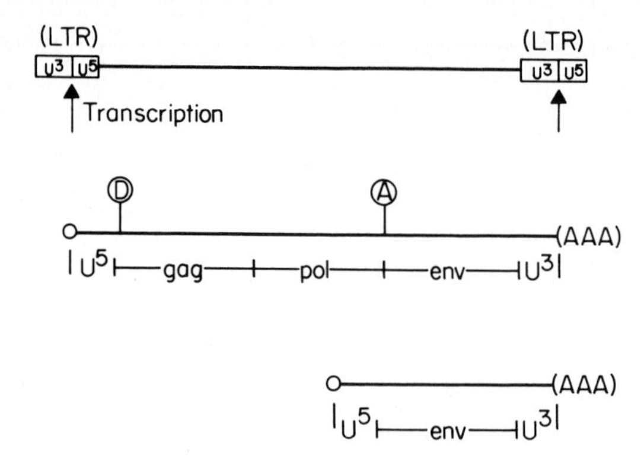

Fig. 1. Diagram of retroviral polynucleotides. The integrated provirus (*top*) is diagrammed. Transcription begins and terminates at the long terminal repeat (*LTR*) units which flank the viral structural genes. The initial 35S transcript (*middle*) contains the full viral genome with viral genes in the order *gag*, *pol*, and *env*. Locations of the splice donor and acceptor sites are diagrammed. Splicing yields the subgenomic 21S RNA (*lower*). Both spliced and unspliced species of RNA are present in nearly equal amounts within the infected cell

product of the viral *pol* gene, to yield a double-stranded DNA copy which integrates into the host chromosome prior to virus production. Transcription of viral RNA then occurs with the integrated provirus as template (Fig. 1).

While viral replication does not necessarily harm host cells, transforming viruses have been identified which induce rapid transformation in tissue culture and tumor formation in animals. These acute transforming viruses have been shown to contain nucleotide sequences similar to host genes (STEHELIN et al. 1976). Some of these genes, termed oncogenes, have recently been found to be mutated in and partially responsible for the formation of tumors (TABIN et al. 1982; REDDY et al. 1982).

2 Microinjection of mRNA

Early microinjection studies identified the functional messenger RNA for envelope glycoprotein by taking advantage of the *env* deficiency of the Bryan strain of Rous sarcoma virus (RSV). Cells infected by this virus, RSV(−) cells, do not release infectious virus unless complemented in envelope glycoprotein function. This complementation was observed when 21S subgenomic mRNA from infected cells was microinjected into RSV(−) cells, but not when the 35S full-genomic RNA was injected (STACEY et al. 1977; Fig. 2). Furthermore, the titer of virus released was roughly proportional to the concentration of 21S mRNA injected (Fig. 3). Since both molecules contained the *env* gene (Fig. 1), it was apparent the gene had to be properly positioned to be translated (VAN ZAANE et al. 1977; PAWSON et al. 1977). Studies have shown the subgenomic 21S mRNA to be formed during splicing (WEISS et al. 1977). It is now clear that splicing is required to position the NH_3 terminus of the envelope glycoprotein gene next to the bulk of *env*-coding sequences (HACKETT et al. 1982). When the 35S molecule was injected into a *gag* and *pol* mutant, the full-genomic molecule served as messenger for these two genes, as has been shown using a cell-free translation system (VON DER HELM and DUESBERG 1975). The 35S viral RNA was also efficiently utilized as genome by virus released from injected cells (STACEY et al. 1977; Fig. 2b).

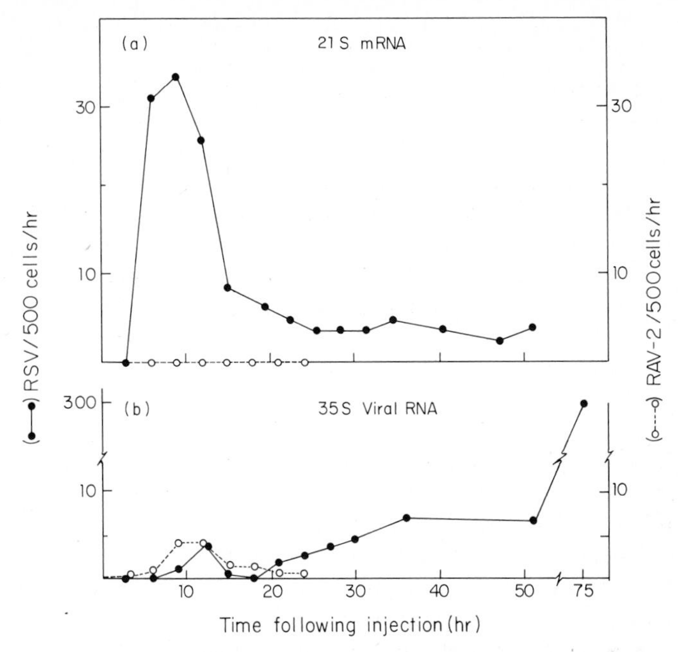

Fig. 2a, b. Virus release following injection of RNA into RSV(−) cells. 21S mRNA from virus-infected cells (*panel A*) or 35S RNA from virus particles (*panel B*) was microinjected into RSV(−) cells. Virus titers released at varying times thereafter are indicated. Note the 3-h delay in virus production following 21S RNA injection and the fact that virus production did not decrease to zero but continued unchanged for up to 50 h. 35S RNA lead to the production of few viruses early after injection, but those which were produced were likely to have packaged the injected RNA to yield RAV-2 virions (*open circles*). These replicating viruses, which were not observed following 21S RNA injections, rapidly spread throughout the injected culture. (From Stacey et al. 1977)

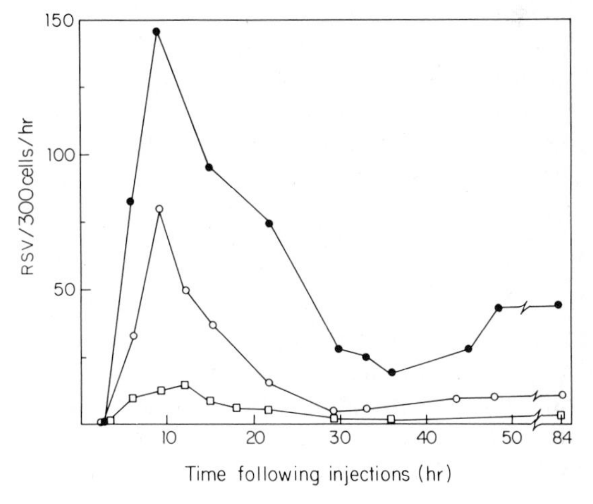

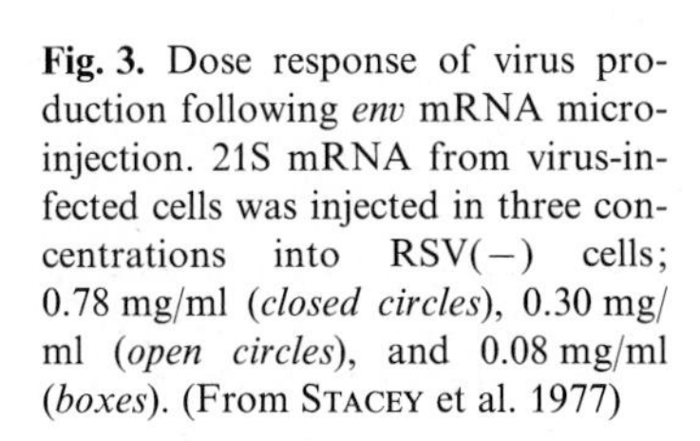

Fig. 3. Dose response of virus production following *env* mRNA microinjection. 21S mRNA from virus-infected cells was injected in three concentrations into RSV(−) cells; 0.78 mg/ml (*closed circles*), 0.30 mg/ml (*open circles*), and 0.08 mg/ml (*boxes*). (From Stacey et al. 1977)

The 35S RNA obtained from virions was also unable to promote virus production following injection into the cytoplasm of RSV(−) cells as was observed with 35S viral RNA from infected cells (Fig. 2). Interestingly, however, when the 35S RNA was microinjected into the nuclei of these cells, efficient virus production resulted. The time course of virus release was similar to that observed following cytoplasmic injection of 21S mRNA (STACEY and HANAFUSA 1978). Apparently, the purified virion RNA had been correctly spliced within the injected nuclei to yield *env* mRNA. This demonstrated, for perhaps the first time, that splicing (CHOW et al. 1977; BERGET et al. 1977) can occur on performed RNA molecules and is localized within the nucleus. It also demonstrated that virion RNA serves not only as genome and messenger for 5′ viral genes, but as nuclear precursor to the 3′ viral *env* gene. Consequently, when 35S virion RNA alone was injected into the nuclei of uninfected chick cells, these cells were able to release fully infectious virus particles (STACEY and HANAFUSA 1978).

3 Studies of Rare Biological Activities

While cell-free systems complemented microinjection studies to clearly identify the primary translational capacity of viral mRNA (VAN ZAANE et al. 1977; PAWSON et al. 1977), the microinjection technique alone has allowed detailed analysis of minor biological functions expressed by these molecules. These activities were observed because living host cells were the subject of analysis. The results obtained are particularly relevant in relation to recent DNA studies of the viral RNA packaging site.

The first indication that 21S viral RNA might have activities in addition to its messenger function come from the time course of viral production following injection into RSV(−) cells. A burst in the release of transforming virus was observed between 3 and 20 h following injection after which virus production decreased to low levels but often did not cease (Figs. 2, 3). Over the next 4 to 35 days, virus production often continued and increased to levels much higher than observed within the first 20 h following injection. The viruses released were unique in that, while they were fully infectious and therefore contained all viral gene products, they did not contain the *env* gene within the genome (STACEY 1980). This gene product appeared, therefore, to be expressed by the cells of the injected culture in a form distinct from the RSV(−) viral genome. In some cultures, however, at various times 4–20 days following injection, a fully replication-competent virus would appear within the culture, presumably due to recombination.

This observation has been explained by assuming that *env* mRNA expresses the function of a genomic molecule, but at reduced efficiency. Accordingly, it was proposed that injected *env* mRNA had been packaged into virus particles released by injected cells. Upon infection of neighboring cells in the RSV(−) culture, this *env* mRNA was then copied into a subgenomic proviral DNA molecule able to direct the production of *env* mRNA. The infected RSV(−) cell would be complemented in virus production by the subgenomic provirus.

The *env* gene would not, however, be present as part of the RSV(−) genome and would not, therefore, be expressed by the virus released unless recombination between the RSV(−) genome and the *env* mRNA took place to generate a new replicating virus.

The model assumes that *env* mRNA must be packaged into virions and must be correctly reverse-transcribed and integrated to form a functional subgenomic provirus. In short, *env* mRNA would be predicted to express all the activities of the genomic RNA, but at reduced efficiency. These assumptions have been confirmed at least in part by the demonstration that long-term, defective virus production requires virus spread within the injected culture (STACEY 1980). Additional confirmation has resulted from analysis of *env* mRNA packaging and the recombinants arising from injected cultures. Finally, approximately one virus in 2000 released by such a culture was shown to express the *env* gene, but in a form distinct from a viral genome. This would suggest that the subgenomic *env* mRNA could be packaged and reverse-transcribed, but with low efficiency as predicted (STACEY 1980).

4 *env* mRNA Packaging

The prediction that *env* mRNA is packaged into the virion was sensitively tested by microinjection as described above. RNA from virus particles was denatured, and fractions of varying size were microinjected into RSV(−) cells. Transforming viruses were actively released, but only following injection of 21S virion RNA (Fig. 4). As before, the major 35S virion species was inactive as *env* mRNA (STACEY 1979). While *env* mRNA activity was apparent within the virion RNA, no physical peak of RNA corresponding to 21S mRNA could be identified. Instead, the 21S region of the virion RNA profile appeared to be composed of a uniform collection of fragmented 35S RNA. Fragmentation studies were, therefore, conducted to demonstrate that the activity observed corresponded to a small amount of genuine *env* mRNA within the virus rather than fragments of 35S virion RNA. Purified 35S virion RNA was fragmented either chemically or enzymatically to yield 21S species. Absolutely no activity was associated

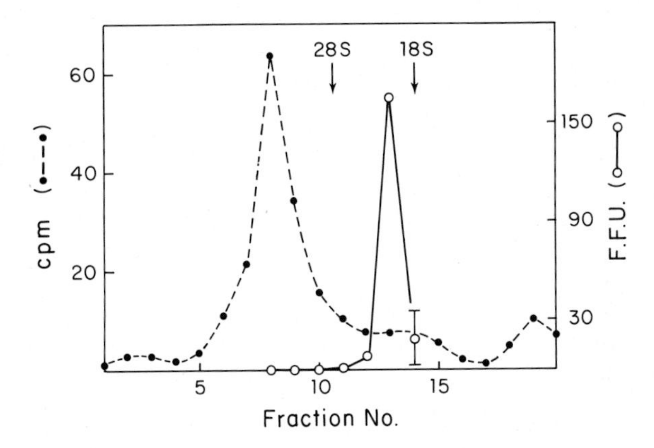

Fig. 4. Injection of denatured virion RNA. Tritiated RNA from virus particles was denatured and sedimented on a sucrose gradient. RNA in the indicated fractions was then injected into RSV(−) cells. The focus-forming units (*FFU*) indicate virus released following injection. (From STACEY 1979)

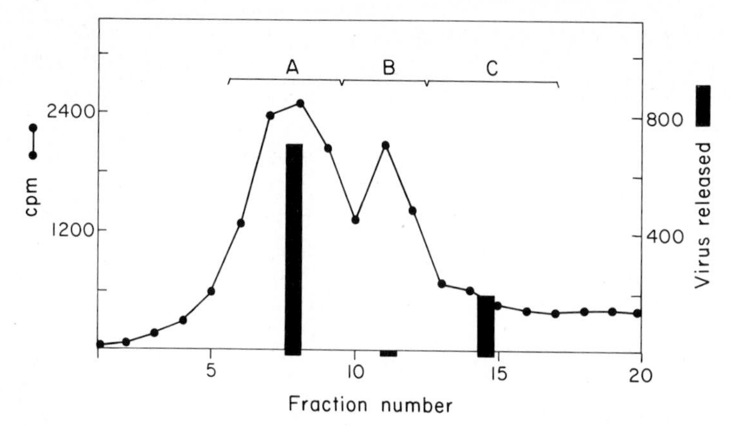

Fig. 5. The *env* mRNA activity of partially denatured virion RNA. Tritiated RNA from virus particles was heated to 54° C to partially denature the high molecular weight complex. After sucrose gradient sedimentation, the 50S–70S RNA (**A**) was injected along with the 35S full-genomic RNA (**B**) and the 21S RNA (**C**). Approximately equal proportions of 35S genomic RNA and 21S *env* mRNA had been released from the virion RNA. (From STACEY 1979)

with the RNA fragments, confirming that the activity identified corresponded to *env* mRNA (STACEY 1979).

These results were not quantitative, so that no comparison of the amount of *env* mRNA packaged could be obtained. Agarose gels of virion RNA, however, do reveal a faint band at 21S which may correspond to packaged subgenomic mRNA (unpublished data). The amount would be only perhaps 1% that of 35S genomic molecules. It is therefore apparent that, while 21S *env* mRNA can be packaged into virions, this occurs infrequently. The inefficient packaging might result either from the reduced size of *env* mRNA or from the elimination of nucleotide sequences specifically recognized by the viral packaging mechanism (IKAWA et al. 1974). Since molecules approximately the size of *env* mRNA have been shown to be efficiently placed in virions (ANDERSON and CHEN 1981), it appears that a sequence required for efficient packaging is localized within the *env* mRNA intron. Use of recombinant DNA analyses to identify this sequence will be described below.

To analyze virion packaged *env* mRNA more carefully, the high molecular weight, 50S–70S virion RNA complex was analyzed. Almost all *env* mRNA activity sedimented with the high molecular-weight RNA as analyzed by RSV(−) injection. The 50S–70S complex was then partially denatured and placed on a sucrose gradient such that approximately 30% of the virion RNA sedimented as 35S subunits. When RNA across this gradient was analyzed, approximately 30% of the *env* mRNA activity had been released from the 50S–70S RNA and sedimented at 21S (Fig. 5; STACEY 1979). The packaged RNA appeared, therefore, to be intimately associated with other RNA in the virion. It is known that the specific association between genomic molecules within the virion is complex but likely to be highly specific (HASELTINE et al. 1977). The point of closest association has been identified using electron microscopic observation

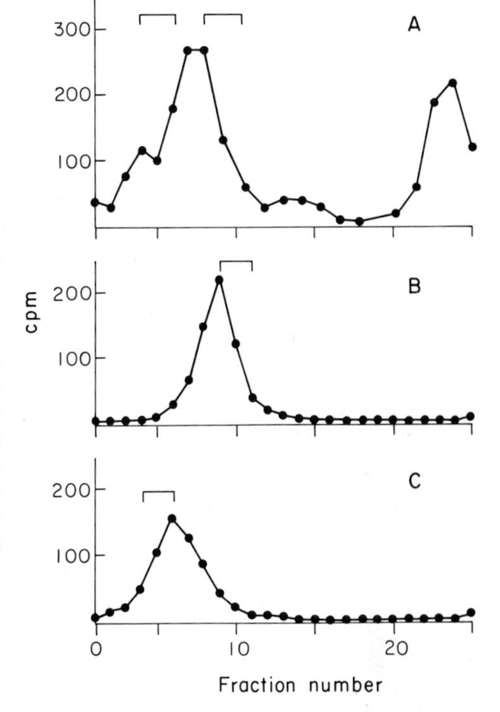

Fig. 6A–C. Association of *env* mRNA with the most rapidly sedimenting virion RNA. Undenatured 50S–70S virion RNA was sedimented (*towards the left, panel A*). The most slowly and most rapidly sedimenting portions of the RNA (indicated by *brackets*) were resedimented (in *panels B* and *C*, respectively). Finally, the most rapidly sedimenting portion of the most rapidly sedimenting RNA fraction and the most slowly sedimenting fraction of the slowly sedimenting RNA were injected into RSV(−) cells and compared, demonstrating that 90% of the *env* mRNA was associated with the most rapidly sedimenting virion RNA. (From STACEY 1979)

of some retroviral RNAs and found to be near the 5′ RNA terminus (CHIEN et al. 1980; MURTI et al. 1981).

The final question regarding *env* mRNA packaging concerns its ability to participate fully in the virion RNA complex to the exclusion of a full-genomic molecule. If a virion packages an *env* mRNA molecule to the exclusion of the full-genomic molecule, the resulting high molecular weight complex would be expected to sediment slightly more slowly than a typical virion RNA complex containing two 35S molecules. If, on the other hand, the *env* mRNA were included in a virion which contains both the virion molecules normally present, the resulting virion complex (in which *env* mRNA is known to be intimately associated) would sediment slightly more rapidly than normal.

To determine the sedimentation characteristics of those few virion RNA complexes which contain *env* mRNA, undenatured virion RNA was sedimented on two cycles of sucrose gradients. The most rapidly sedimenting RNA from the initial gradient was resedimented and the most rapidly sedimenting RNA of the second gradient was collected. Similarly, the most slowly sedimenting RNA purified from two successive gradients was collected. These RNA were then denatured and microinjected into RSV(−) cells. The most rapidly sedimenting RNA preparation contained approximately tenfold more *env* activity than the more slowly sedimenting species (STACEY 1979; Fig. 6). This tedious experiment was repeated numerous times to ensure that the result was correct.

Association of *env* mRNA with a larger than normal virion RNA complex provides important insight into the nature of subgenomic RNA packaging.

It is clear that *env* mRNA is added to a normal complex containing two full-genomic molecules. This indicates that *env* mRNA did not participate in the normal packaging mechanism by which two and only two molecules are placed into the virion. Instead, the *env* mRNA appears to have been carried along nonspecifically into the virion. This may suggest that multiple levels of recognition exist in packaging. To fully participate, a molecule must contain the full genome. The subgenomic molecule must have lost part of this recognition system during splicing but may, nevertheless, still possess sequences recognized much less efficiently by the packaging system. Alternately, *env* mRNA may be packaged with no greater efficiency than nonviral molecules, such as globin mRNA, but is recognized in the analyses above because of the sensitive microinjection assay employed.

Regardless of the packaging mechanism, it is clear that once within the virion, *env* mRNA can be converted into a fully active subgenomic provirus. A rough indication of the frequency of this event may be obtained by comparison of the proportion of *env* mRNA within the virion to the proportion of virus released by 21S *env* mRNA-injected RSV(−) cultures which are able to express *env* activity in the absence of a viral genome containing *env*. There appears to be perhaps 1% *env* mRNA in the virion, and approximately 0.1% of viruses released by *env* mRNA-injected cultures express a subgenomic *env* provirus. This fact may be of importance in models of reverse-transcription. In addition, the replicating viruses which spontaneously appeared several days following *env* mRNA injection into RSV(−) cells were conclusively shown to be generated by recombination between *env* mRNA and the RSV(−) genome (Wang and Stacey 1982). It would therefore appear that the major distinction between full-genomic viral RNA and subgenomic RNA is packaging efficiency. Once within the virion, the subgenomic molecule behaves as a genomic molecule. This would localize functions involved in replication to those regions common to full and subgenomic viral RNAs. The signal recognized in packaging, on the other hand, is lost during splicing. This assumption provides the basis for more detailed analyses (described below) involving DNA injection.

5 DNA Injections

While the studies described above were aided by the sensitivity of the viral genetic analyses employed, they were limited by the number of naturally occurring viral RNAs available for study. More detailed structure and function correlation required the type of molecular manipulation which can be performed on DNA molecules. At first, it was not known, however, whether injected retroviral DNA would be biologically active (see Capecchi 1980).

To determine the activity of injected DNA, a clone of RSV was obtained. This clone had been generated by cleavage of a circular viral DNA molecule obtained from an infected cell. The linearized molecule was then inserted into a bacterial vector. The molecule, therefore, contained all viral genetic sequences but the viral-specific region was a linear permutation of the viral genome, with parts of the *env* gene at each terminus (Cullen et al. 1982; Fig. 7A). For this

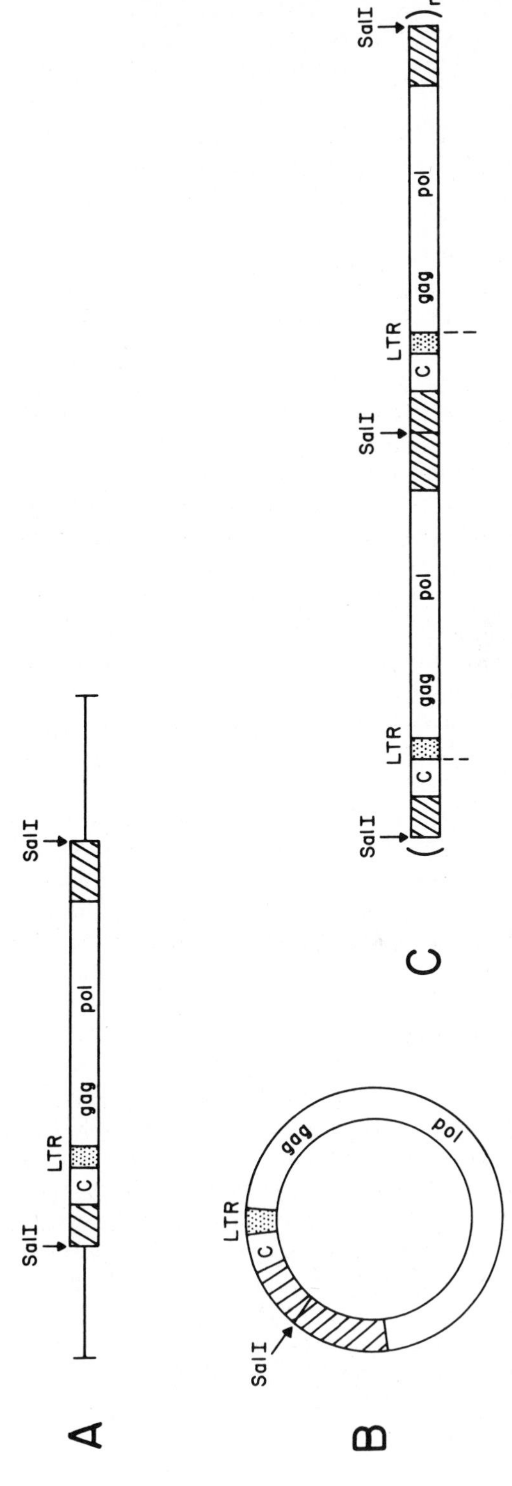

Fig. 7A–C. DNA clone and ligation. **A** The permuted retroviral plasmid is diagrammed, with the *wide bar* representing viral-specific sequences and the *env* gene represented by *crosshatching*. After cleavage of the plasmid with *Sal*I, the viral-specific DNA is released. Within the cell, this DNA can be ligated to reconstitute an intact *env* gene either by circle formation (**B**) or ligation to form a molecular concatomer (**C**)

DNA to be active in complementing injected RSV(−) cells, the DNA would not only have to be transcribed, but would also have to be ligated to reconstitute *env* (KOPCHICK et al. 1981b). It was therefore perhaps surprising not only that the injected DNA complemented RSV(−) cells in virus production, but that viruses were released beginning 3 h following injection (KOPCHICK et al. 1981b). This same lag was observed following *env* mRNA injection. Not only had transcription and ligation occurred, but these had taken place very soon after the injection. Virus production continued for up to 9 days following these injections and was much more efficient following injection directly into the nucleus. When a full-genomic clone was injected into uninfected cells, virus production was observed, with a peak near 24 h following injection.

The ability of injected DNA molecules to be efficiently ligated by the cell is not only interesting in itself but might provide great experimental flexibility. A careful analysis of intracellular ligation was therefore undertaken (KOPCHICK et al. 1981a). The results were compared with those obtained in the same cells by diethylaminoethanol (DEAE) dextran mediated transfection. It was conclusively demonstrated that within avian cells injected as well as transfected DNAs are ligated rapidly and with great efficiency. Ligation of injected DNA may involve only a single molecule to generate a circle, or it may join many molecules together to form a high molecular weight complex (KOPCHICK and STACEY 1983; Fig. 7B, C).

Injected DNA was not only efficiently ligated, but the ligation was highly restrictive. While blunt termini were joined, single-stranded termini generated by restriction endonuclease treatment were joined only to complementary termini. In such a case, mutations were rarely produced. Indeed, an entire viral genome has been efficiently regenerated within a cell which received an injection of viral genomic DNA in four separate restriction fragments. In contrast, DNA transfected into the same cells was joined regardless of molecular topology. The joining was normally accompanied by the introduction of limited deletions at the molecular termini (KOPCHICK and STACEY 1984).

With the demonstration that DNA can be efficiently transcribed following microinjection comes a new approach for studying viral structure and function relationships. In recent work, analyses have been made of the requirements for viral splicing, transcription, the primer binding site, and the packaging sequence. It has been demonstrated that large deletions and insertions of foreign DNA can be made into the intron of the *env* gene without abolishing proper transcription and splicing (CULLEN et al. 1982; PUGATSCH and STACEY 1982). The effect of transcriptional promoters and terminators located within the viral genome upon viral transcription and splicing have been studied (STACEY, KOPCHICK, and KAHN, unpublished data). The activity of these sequences within the environment of the viral genome has also been tested. In addition, the location within the viral genome at which a foreign gene can be maximally expressed has also been determined (MULCAHY, PESTKA, and STACEY, unpublished data). The analysis of greatest relevance to the present discussion, however, relates to those sequences required for efficient virion packaging.

6 DNA Studies of Packaging

On the basis of *env* mRNA studies, it was apparent that a sequence required for efficient virion packaging is localized within the *env* mRNA intron. This conclusion came not only from the observations that little *env* mRNA is present within the virion, but also from the observation that those *env* mRNA molecules which become packaged do so in an aberrant manner. On the other hand, previous studies suggest that activities other than virion packaging are expressed by *env* mRNA. Sequences required for these functions are apparently not removed with the *env* intron.

On the basis of these considerations, a microinjection assay was developed to test the efficiency with which the transcript of an injected viral DNA could be packaged. In this assay cloned viral DNA was injected into the nuclei of RSV(−) cells. Transcription would yield genomic RNA which might either be exported directly to the cytoplasm or spliced to yield *env* mRNA. The *env* mRNA formed would ensure that infectious virus would be released from the injected cells. These virions would then be able to package either RSV(−) genomic molecules or the full-genomic transcripts of the injected DNA. The ratio of RSV(−) molecules to DNA transcripts packaged could then be determined genetically. This ratio would indicate the efficiency with which the transcript was packaged (Pugatsch and Stacey 1983).

Attempts to identify the packaging site initially involved large deletions introduced near the middle of the 4.7 kbp *env* intron. Deletions up to 3.0 kbp were tested and found to be unaltered in packaging efficiency compared with normal viral DNA. In contrast, deletions of 40–70 bp near the *Sst*II restriction site reduced packaging efficiency nearly 100-fold. The *Sst*II site is located approximately 150 bp downstream of the splice donor site for *env* mRNA (Fig. 8).

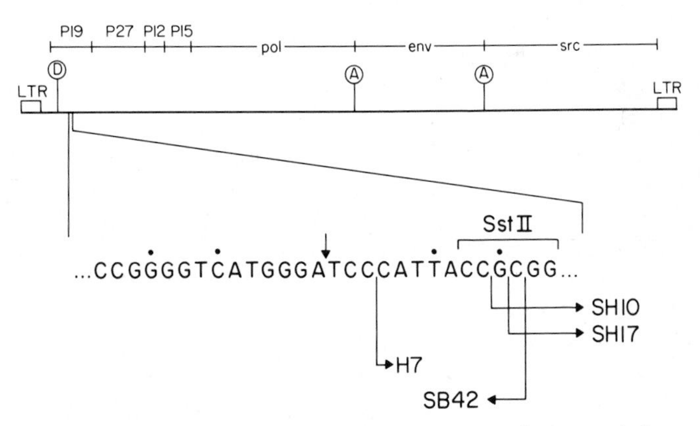

Fig. 8. Localization of sequences necessary for packaging. The viral genome is diagrammed along with the localization of viral peptide-coding sequences and splice donor and acceptor sites. The 28-bp palendrome containing the *Sst*II site which is located just downstream of the splice donor site is diagrammed along with DNA mutants introduced into this region. Of the mutants diagrammed, only SH17 was active in packaging. (From Pugatsch and Stacey 1983)

To critically identify sequences required in packaging, other deletions were introduced. Deletions extending from the *Sst*II site in the 3′ direction were tested. Two of these deletions differed by only 2 bp yet one, SH17, retained wild-type packaging efficiency while the other, SH10, displayed the low packaging of other deficient mutants (Fig. 8). Apparently, the 3′ terminus of a sequence required for packaging corresponds to the region near the *Sst*II site (PUGATSCH and STACEY 1983). Interestingly, the *Sst*II site is the 3′ terminus of a self-complementary region of 28 bp termed the dimer linkerage (Fig. 8). It is interesting that this sequence appears to be involved both in packaging of viral RNA molecules and in allowing them to associate within the virion. The fact that *env* mRNA lacks this sequence might explain why those few *env* mRNA molecules which do become packaged do not displace genomic molecules from the virion RNA complex. Genomic regions required for packaging of other types of virus have also been identified to be downstream of the splice donor site (WATANABE and TEMIN 1982; MANN et al. 1983). In addition, other areas of the viral genome have been implicated in virion packaging (SHANK and LINIAL 1980; SORGE et al. 1983).

Attempts to introduce subtle alterations into the 28 bp palendrome have as yet failed to provide further information on the sequence requirements for packaging. Indeed, it appears that the packaging mechanism may be complicated and may involve additional genomic sequences. When specific base changes were introduced near the *Sst*II restriction site sufficient to desrupt base pairing in the 28bp palendrome, packaging was often unaffected (DE GUIDICIBUS and STACEY, in preparation). The continued investigation into viral packaging promises to be an interesting one which may hold important clues into the correlation between RNA secondary structure and recognition sequences. As in the past, the technique of microinjection should prove useful in these studies.

7 Protein Injections

Little has been said about protein injections, yet polynucleotides within a living cell are always associated with and modulated by proteins. There is no question that future microinjection studies will explore the behavior of protein nucleic acid complexes. In addition, proteins themselves may be tested for their effect upon cells.

As an example of the type of studies which can be performed, the protein product of the Balb murine sarcoma virus transforming gene (*ras*) was injected into NIH 3T3 cells. Recipient cells underwent a transient morphological transformation (Fig. 9). In addition, the injected cells were efficiently induced to enter the S-phase of the cell cycle by the injected protein (STACEY and KUNG 1984). On the other hand, when monoclonal antibody to cellular *ras* proteins was injected into fibroblasts initiation of S-phase was inhibited (MULCAHY et al. 1985). These data indicate that *ras* proteins are sufficient and necessary for fibroblast cell division and must therefore be intimately involved in control of proliferation.

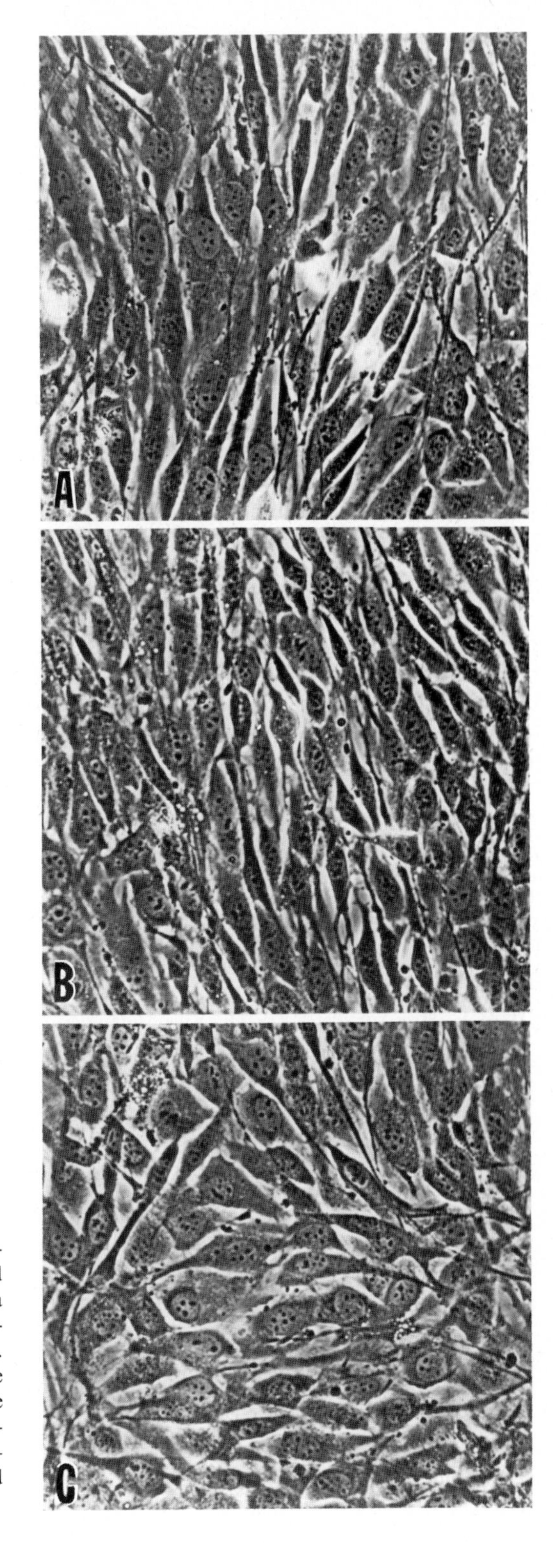

Fig. 9A–C. Cells injected with *ras* protein. These NIH 3T3 cells were injected with viral *ras* protein from Balb sarcoma virus (**B**), with the normal human *ras* protein (**A**), or an unrelated protein (**C**). Note that the viral *ras*-injected cells are more elongated and refractile than the control cells. The normal *ras* protein-injected cells show an intermediate morphology (unpublished data, STACEY and KUNG)

8 Summary

The technique of microinjection along with viral genetics and molecular biology has proven useful in the correlation of retroviral polynucleotide structure with function. The advantage of this technique is the involvement of living cells where rare activities may be observed and where properties of living cells can be assayed. Future studies involving recombinant DNA molecules and the association of proteins with nucleic acids promise to yield additional insight into the nucleotide sequences involved in the expression of viral activities.

References

Anderson SM, Chen JH (1981) In vitro translation of avian myeloblastosis virus RNA. J Virol 40:107–117

Berget SM, Moore C, Sharp PA (1977) Spliced segments of the 5′ terminus of adenovirus 2 late mRNA. Proc Natl Acad Sci USA 74:3171–3175

Bishop JM (1978) Retroviruses. Annu Rev Biochem 47:35–88

Capecchi MR (1980) High efficiency transformation by direct microinjection of DNA into cultured mammalian cells. Cell 22:479–488

Chien, Y-H, Junghans RP, Davidson W (1980) Electron microscopic analysis of the structure of RNA tumor virus nucleic acids. In: Stephenson JR (ed) Molecular biology of RNA tumor viruses. Academic, New York, pp 395–446

Chow LT, Gelinas RE, Broker TR, Roberts RJ (1977) An amazing sequence arrangement at the 5′ ends of adenovirus 2 messenger RNA. Cell 12:1–8

Coffin JM (1979) Structure, replication and recombination of retrovirus genome: some unifying hypotheses. J Gen Virol 42:1–26

Cullen BR, Kopchick JJ, Stacey DW (1982) Effects of intron size on splicing efficiency in retroviral transcripts. Nucleic Acids Res 10:6177–6189

Hackett PB, Swanstrom R, Varmus HE, Bishop JM (1982) The leader sequence of the subgenomic mRNA's of Rous sarcoma virus is approximately 390 nucleotides. J Virol 41:527–534

Haseltine WA, Maxam AM, Gilbert W (1977) Rous sarcoma virus genome is terminally redundant: the 5′ sequence. Proc Natl Acad Sci USA 74:989–993

Hayward WS (1977) Size and genetic content of viral RNAs in avian oncovirus-infected cells. J Virol 24:47–63

Ikawa Y, Ross J, Leder P (1974) An association between globin messenger RNA and 60S RNA derived from friend leukemia virus. Proc Natl Acad Sci USA 71:1154–1158

Kopchick JJ, Stacey DW (1983) Selective ligation of DNA molecules following microinjection. J Biol Chem 258:11528–11536

Kopchick JJ, Stacey DW (1984) Differences in intracellular ligation after microinjection and transfection. Mol Cell Biol 4:240–246

Kopchick JJ, Cullen BR, Stacey DW (1981a) Rapid analysis of small nucleic acid samples by gel electrophoresis. Anal Biochem 115:419–423

Kopchick JJ, Ju G, Skalka AM, Stacey DW (1981b) Biological activity of cloned retroviral DNA in microinjected cells. Proc Natl Acad Sci USA 78:4383–4387

Mann R, Mulligan RC, Baltimore D (1983) Construction of a retroviral packaging mutant and its use to produce helper-free defective retrovirus. Cell 33:153–159

Mulcahy LS, Smith MR, Stacey DW (1985) Requirement for *ras* protooncogene function during serum-stimulated growth of NIH 3T3 cells. Nature 313:241–243

Murti KG, Bondurant M, Tereba A (1981) Secondary structural features in the 70S RNA of Moloney murine leukemia and Rous sarcoma viruses as observed by electron microscopy. J Virol 37:411–419

Pawson T, Harvey R, Smith AE (1977) The size of Rous sarcoma virus mRNAs active in cell-free translation. Nature 268:416–420

Pugatsch T, Stacey DW (1982) Analysis by microinjection of the biological effects of site-directed mutogenesis in cloned avian leukosis viral DNAs. J Virol 43:503–510
Pugatsch T, Stacey DW (1983) Identification of a sequence likely to be required for avian retroviral packaging. Virol 128:505–511
Reddy EP, Reynold RK, Santos E, Barbacid M (1982) A point mutation is responsible for the aquisition of transforming properties by the T24 bladder carcinoma oncogene. Nature 300:149–152
Shank PR, Linial M (1980) Avian oncovirus mutant (SE21Q16) deficient in genomic RNA: characterization of a deletion in the provirus. J Virol 36:450–456
Sorge J, Ricci W, Hughes SH (1983) *Cis*-acting RNA packaging locus in the 115-nucleotide direct repeat of Rous sarcoma virus. J Virol 48:667–675
Stacey DW (1979) Messenger activity of virion RNA for avian leukosis viral envelope glycoprotein. J Virol 29:949–956
Stacey DW (1980) Expression of a subgenomic retroviral messenger RNA. Cell 21:811–820
Stacey DW, Hanafusa H (1978) Nuclear conversion of microinjected avian leukosis virion RNA into an envelope-glycoprotein messenger. Nature 273:779–782
Stacey DW, Kung H-F (1984) Transformation of NIH 3T3 cells by microinjection of H-*ras* p21 protein. Nature 310:508–511
Stacey DW, Allfrey VG, Hanafusa H (1977) Microinjection analysis of envelope glycoprotein messenger activities of avian leukosis viral RNAs. Proc Natl Acad Sci USA 74:1614–1618
Stehelin D, Varmus HE, Bishop JM, Vogt PK (1976) DNA related to the transforming gene(s) of avian sarcoma viruses is present in normal avian DNA. Nature 260:170–173
Tabin CJ, Bradley SM, Bargmann CI, Weinberg RA, Papagcorge AG, Scolnick EM, Dhar R, Lowy DR, Chang EH (1982) Mechanism of activation of a human oncogene. Nature 300:143–149
Van Zaane D, Gielkens ALJ, Hesselink WG, Bloemers HPJ (1977) Identification of Rauscher murine leukemia virus-specific mRNAs for the synthesis of *gag*- and *env*-gene products. Proc Natl Acad Sci USA 74:1855–1859
Von der Helm K, Duesberg PH (1975) Translation of Rous sarcoma virus RNA in a cell-free system from Krebs II cells. Proc Natl Acad Sci USA 72:614–618
Wang L-H, Stacey DW (1982) Participation of subgenomic retroviral mRNAs in recombination. J Virol 41:919–930
Watanabe S, Temin HM (1982) Encapsidation sequences for spleen necrosis virus, an avian retrovirus, are between the 5′ long terminal repeat and the start of the *gag* gene. Proc Natl Acad Sci USA 79:5986–5990
Weiss S, Varmus HE, Bishop JM (1977) The size and genetic composition of virus-specific RNAs in the cytoplasm of cells producing avian sarcoma leukosis viruses. Cell 12:983–992

The Viral Tyrosine Protein Kinases

B.M. Sefton

1 Introduction 40

2 Cellular Transformation 41
2.1 Tyrosine Phosphorylation and Cellular Transformation 42
2.2 Phosphatidylinositol Phosphorylation 43

3 *src* 43
3.1 Subcellular Location of $p60^{src}$ 44
3.2 Structural Domains 44
3.3 Myristylation of $p60^{src}$ 45
3.4 Sites of Phosphorylation 45
3.5 Substrates of $p60^{src}$ 46
3.6 $p60^{c\text{-}src}$ 47

4 *abl* 48
4.1 Structure of the Abelson Virus Transforming Protein 48
4.2 Subcellular Location of $P160^{abl}$ 50
4.3 Sites of Phosphorylation 50
4.4 Polypeptide Substrates 51
4.5 c-*abl* 51

5 *erb*B 52
5.1 Structure of the *erb*B Protein 52
5.2 Subcellular Location of the Polypeptide Products of *erb*B 53
5.3 Sites of Phosphorylation 53
5.4 Polypeptide Substrates 53
5.5 The Polypeptide Product of the c-*erb*B Gene 54

6 *fps* 55
6.1 Structure of the Transforming Protein of FSV 55
6.2 Subcellular Location of $P140^{fps}$ 56
6.3 Sites of Phosphorylation 56
6.4 Polypeptide Substrates 57
6.5 The c-*fps* Product 57

7 *yes* 57
7.1 Structure of the Viral Transforming Protein 57
7.2 Subcellular Location of $P90^{yes}$ 58
7.3 Sites of Phosphorylation 58
7.4 Polypeptide Substrates 58
7.5 The Polypeptide Product of c-*yes* 59

8 *fgr* 59
8.1 Subcellular Location of $P70^{fgr}$ 59
8.2 Sites of Phosphorylation 60
8.3 Polypeptide Substrates 60

Molecular Biology and Virology Laboratory, The Salk Institute, P.O. Box 85800, San Diego, CA 92138, USA

Current Topics in Microbiology and Immunology, Vol. 123

9 *ros* 60
9.1 Subcellular Location of P68^{ros} 60
9.2 Sites of Phosphorylation 61
9.3 Polypeptide Substrates 61
9.4 c-*ros* 61
10 *fms* 62
10.1 Phosphorylation 62
10.2 c-*fms* 62
11 General Properties of the Viral Tyrosine Protein Kinases 63
12 The Evolution and Divergence of the Genes for the Cellular Tyrosine Protein Kinases 63
13 The Future 64
References 64

1 Introduction

The acutely transforming retroviruses have revealed the existence of more than twenty different genes with oncogenic potential. Seven of these – *src*, *abl*, *yes*, *erb*B, *fps* (*fes*), *fgr*, *fms*, and *ros* – encode proteins with intrinsic tyrosine protein kinase activity. The inappropriate phosphorylation of cellular proteins is likely to play a central and crucial role in the malignant transformation of cells in which these genes are expressed.

Invariably, acutely transforming retroviruses have been found to have arisen as a result of the transduction of cellular proto-oncogenes by a relatively benign leukosis virus (BISHOP 1983). The viral oncogenes are therefore largely, and in some cases, wholly, cellular in origin. To distinguish between the viral oncogene and the protooncogene resident in the cellular genome, the terms v-*onc* and c-*onc* are used, respectively.

Table 1. Viral oncogenes and cellular tyrosine protein kinases

Viral oncogene	Product of cellular homologue
src	p60$^{c\text{-}src}$
abl	P150$^{c\text{-}abl}$
*erb*B	Epidermal growth factor receptor
fps	NCP98$^{c\text{-}fps}$
yes	?
fgr	?
ros	?
fms	colony-stimulating factor I receptor
?	Platelet-derived growth factor receptor
?	Somatomedin C receptor
?	Insulin receptor
?	p56
?	p75

p56 is a cellular tyrosine protein kinase identified in murine T cells and expressed at high levels in two Moloney leukemia virus-induced thymoma cell lines (CASNELLIE et al. 1984; VORONOVA et al. 1985); p75 is a cytosolic tyrosine protein kinase purified from rat liver (WONG and GOLDBERG 1984)

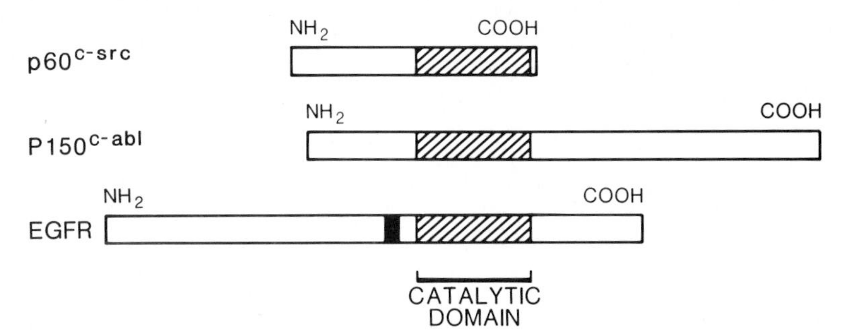

Fig. 1. Comparison of p60^{c-src}, P150^{c-abl}, and the EGF receptor, the cellular homologues of the *src*, *abl*, and *erb*B oncogenes, respectively. Each protein is drawn to scale. The sequences of p60^{c-src} (TAKEYA and HANAFUSA 1983) and the EGF receptor (ULLRICH et al. 1984), have been determined directly. The probable sequence of the c-*abl* product has been deduced from the known sequence of the v-*abl* gene (REDDY et al. 1983) and the structure of c-*abl* mRNA (WANG et al. 1984). The proteins are positioned so that their catalytic domains are aligned. The transmembrane domain of the EGF receptor is indicated in *black*

The diversity of viral oncogenes encoding tyrosine protein kinases reflects the diversity of cellular genes encoding such enzymes. Avian and mammalian genomes contain a minimum of 12 chromosomal loci which code for tyrosine protein kinases (Table 1; SEFTON and HUNTER 1984). The deduced amino acid sequences of the viral tyrosine protein kinases show the genes for this group of proteins to comprise a divergent, evolutionarily related gene family. The large number of genes for this class of enzymes does not have as its basis simple gene amplification, such as the process which gave rise to the highly related globin gene family. p60^{c-src}, P150^{c-abl}, and the receptor for the epidermal growth factor, the products of the cellular homologues of the *src*, *abl*, and *erb*B genes, respectively, are very different in structure, save for homologous catalytic domains (Fig. 1; TAKEYA and HANAFUSA 1983; WANG et al. 1984; ULLRICH et al. 1984). The distinguishing properties of the viral tyrosine protein kinases is the subject of this review.

2 Cellular Transformation

Transformed fibroblasts differ from their normal counterparts in a number of significant respects: (a) They have an altered morphology; (b) they adhere to their substrata poorly; (c) they are no longer sensitive to normal growth regulatory signals; (d) they metabolize glucose aberrantly; and (e) they exhibit an altered pattern of gene expression. The phosphorylation of cellular proteins by viral tyrosine protein kinases most probably leads, either directly or indirectly, to each of these changes. This argument can be made most persuasively in the case of the loss of growth control. At least three peptides with mitogenic activity, epidermal growth factor (EGF), platelet-derived growth factor, and somatomedin C (insulin-like growth factor I) stimulate tyrosine protein kinase activity when they bind to their cell-surface receptors (USHIRO and COHEN 1980;

Table 2. Cellular substrates of viral tyrosine protein kinases

	Size	Subcellular location
Structural proteins		
Vinculin	115K	Adhesion plaques
p81	81K	Microvilli
p36	36K	Cytoplasmic face of plasma membrane
Glycolytic enzymes		
Enolase	46K	Cytosol
Lactate dehydrogenase	35K	Cytosol
Phosphoglycerate mutase	29K	Cytosol
Other proteins		
p50	50K	Cytosol
p42	42K	Cytosol

EK et al. 1982; NISHIMURA et al. 1982; RUBIN et al. 1983; JACOBS et al. 1983). The viral tyrosine protein kinases may therefore induce unrestrained cell replication through the chronic phosphorylation of one or more of the normal substrates of the growth factor receptors.

2.1 Tyrosine Phosphorylation and Cellular Transformation

A prerequisite for understanding the biochemistry of cellular transformation by viruses encoding tyrosine protein kinases is identification of their cellular substrates. Eight presumptive substrates have been identified (Table 2). Three of these are structural proteins. One is vinculin, a 115000 dalton cytoskeletal protein associated with areas of the cytoplasmic face of the plasma membrane at which actin-containing microfilaments are anchored (GEIGER 1979; SEFTON et al. 1981b). Another is p81, an 81000 dalton polypeptide present in microvilli (BRETSCHER 1983; COOPER and HUNTER 1981c; HUNTER and COOPER 1983; K. GOULD 1985). The third is p36, a 34000 to 39000 dalton protein component of the filamentous cortex located beneath the plasma membrane (RADKE et al. 1980, 1983; ERIKSON and ERIKSON 1980; COOPER and HUNTER 1981a, 1982; COURTNEIDGE et al. 1983; NIGG et al. 1983). Three substrates are the glycolytic enzymes enolase, lactate dehydrogenase, and phosphoglycerate mutase (COOPER et al. 1983). Another, p42, is a cytosolic protein which is also the substrate of tyrosine protein kinases stimulated by mitogenic agents (COOPER and HUNTER 1981a; COOPER et al. 1982, 1984; NAKAMURA et al. 1983; BISHOP et al. 1983; GILMORE and MARTIN 1983). The last, termed p50, is a 50000-dalton polypeptide of unknown function (HUNTER and SEFTON 1980). It has, however, the interesting property of binding tightly and specifically to three of the viral tyrosine protein kinases (BRUGGE et al. 1981; ADKINS et al. 1982; LIPSICH et al. 1982). To a first approximation, all of the viral tyrosine kinases induce the phosphorylation of the same spectrum of cellular proteins. There are, however, several notable exceptions to this generalization and they will be discussed below.

The question of whether the phosphorylation of any of these particular cellular proteins plays an important role in the transformation of cells remains unanswered. It may be that the crucial substrates of the viral tyrosine protein kinases have yet to be identified. A thorough and critical discussion of the phosphorylation of cellular substrates can be found in the chapter by COOPER and HUNTER (1984) in Vol. 107 of this series.

2.2 Phosphatidylinositol Phosphorylation

The biochemistry of cellular transformation induced by the viral tyrosine protein kinases may have another level of complexity. It has been reported recently that the products of both the *src* and *ros* genes, $p60^{src}$ and $P68^{ros}$, can phosphorylate phosphatidylinositol in vitro (MACARA et al. 1984; SUGIMOTO et al. 1984). If this reaction occurs at a significant rate in infected cells it could have important physiological consequences.

The phosphorylation of phosphatidylinositol produces phosphatidylinositol 4,5 bisphosphate, an excellent substrate for phospholipase C. Because the hydrolysis of phosphatidylinositol 4,5 bisphosphate by phospholipase C yields inositol triphosphate and diacylglycerol, increased phosphorylation of phosphatidylinositol could lead to increased production of diacylglycerol. This is of note because diacylglycerol stimulates the activity of the serine-threonine-specific protein kinase C (NISHIZUKA 1983), an enzyme which may be capable of inducing a partially transformed cell phenotype.

The phorbol diester tumor promoters such as tetradecanoyl phorbol acetate (TPA) induce in some cells a reversible form of transformation which is not totally dissimilar to that seen in virally infected cells (BLUMBERG et al. 1976; BISSELL et al. 1979; RIFKIN et al. 1979). These promoters, like diacylglycerol, are able to stimulate C-kinase in vitro (CASTAGNA et al. 1982). Their effects on cells may result therefore from stimulation of C-kinase. Similarly, stimulation of the rate of production of diacylglycerol should activate protein kinase C and this could well lead to changes in cell shape, growth rate, and metabolism similar to those observed transiently with tumor promoters. Experimental support for this notion has in fact already been obtained. ROZENGURT et al. (1984) have shown that addition of diacylglycerol to quiescent 3T3 cells can reverse the growth inhibition induced by serum starvation. Determination of whether $p60^{src}$ and $P68^{ros}$ carry out this reaction in infected cells and whether this leads to a significant increase in the level of diacylglycerol is obviously of importance.

3 *src*

The first viral oncogene characterized in detail (STEHELIN et al. 1976) and the first found to encode a protein kinase (COLLETT and ERIKSON 1978; LEVINSON et al. 1978) with a specificity for tyrosine (HUNTER and SEFTON 1980) is that present in Rous sarcoma virus (RSV), *src*. This virus was isolated in the field from a chicken and is now known to have arisen from the acquisition by a weakly oncogenic avian leukosis virus of the chicken c-*src* gene. The product of the v-*src* gene is a 60000 phosphoprotein termed variously p60^{src}, pp60^{src}, or pp60$^{v\text{-}src}$ (Purchio et al. 1978). Unlike the majority of viral transforming proteins, it is encoded entirely by acquired cellular genetic sequences.

3.1 Subcellular Location of p60^{src}

p60^{src} is found in the cytoplasm of infected cells where a majority, often more than 80%, is present on the cytoplasmic face of the plasma membrane (Kreuger et al. 1980a; Courtneidge et al. 1980; Krzyzek et al. 1980). In the plasma membrane, it is found concentrated at cell:cell junctions and at the adhesion plaques which both anchor actin-containing microfilaments to the plasma membrane and are the sites of closest contact between the lower surface of the cell and the substratum (Willingham et al. 1979; Rohrschneider 1980). Under some circumstances, significant amounts of the protein are found associated with intracellular membranes – specifically those of the rough endoplasmic reticulum and the nucleus (Kreuger et al. 1980b; Garber et al. 1982; Resh and Erikson 1985). Some p60^{src} is also free in the cytosol. This population is invariably recovered as part of a trimolecular complex consisting of p60^{src}, a 90000 dalton heat-shock protein, termed hsp90, and the 50000-dalton presumptive substrate of p60^{src}, p50 (Brugge et al. 1981). The significance of this subpopulation of p60^{src} is an enigma. There is no convincing evidence that p60^{src} is present at the cell surface or in any cellular organelle. A thorough discussion of the distribution of p60^{src} in transformed cells can be found in Kreuger et al. (1983) in Vol. 107 of this series.

3.2 Structural Domains

p60^{src} is composed of two functional domains (Fig. 2). Limited proteolysis has shown that the carboxy terminal 30000 daltons contains the totality of the catalytic domain (Levinson et al. 1981). There is direct evidence that lysine 295, which is located approximately in the middle of p60^{src}, forms part of the ATP-binding site (Kamps et al. 1984). The amino terminal 8000 to 13000 daltons serves to anchor the protein to cellular membranes (Kreuger et al. 1980a; Levinson et al. 1981). Both domains are essential for cellular transformation. Mutations within 15 residues of the amino terminus render the protein unable to associate stably with cellular membranes and abolish transformation (see below) without affecting kinase activity (Cross et al. 1984; Kamps et al. 1985). Conversely, point mutations at several sites within the carboxy terminal half of the molecule eliminate both protein kinase activity and cellular transformation (Bryant and Parsons 1983, 1984).

In contrast, much of the protein located between these two domains appears not to be essential for either kinase activity or cellular transformation. A deletion

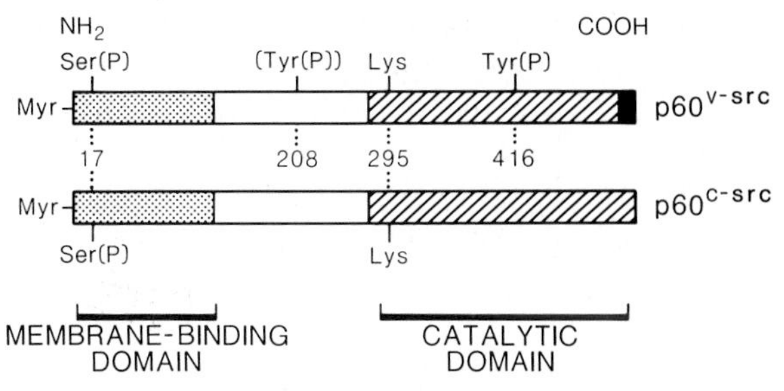

Fig. 2. Comparison of p60$^{v\text{-}src}$ and p60$^{c\text{-}src}$. Indicated are the amino terminal myristyl moiety, the phosphorylated serine residue at position 17, the phosphorylated tyrosine at position 208 (found only in the p60$^{v\text{-}src}$ protein of Prague strain *RSV*), lysine 295 (present in the ATP-binding site), tyrosine 416 (phosphorylated in p60$^{v\text{-}src}$ but not in p60$^{c\text{-}src}$), and the unusual carboxy terminus of the viral polypeptide

which removes residues 173–227 yields a protein which is temperature sensitive in its ability to transform cells (BRYANT and PARSONS 1982). Another which affects residues 135–232 yields a species of p60^{src} which transforms cells but induces an elongated, rather than a fully rounded transformed cellular morphology (KITAMURA and YOSHIDA 1983).

3.3 *Myristylation of p60^{src}*

p60^{src} undergoes an unusual form of posttranslational modification. The rare 14-carbon saturated fatty acid, myristic acid, is attached to the alpha-amino group of the amino terminal glycine residue (BUSS and SEFTON 1985; SCHULZ et al. 1985). This modification plays a crucial role in the association of p60^{src} with cellular membranes. Mutations which convert the amino terminal glycine to alanine prevent myristylation and yield a cytosolic species of p60^{src} which induces only slight cellular transformation (KAMPS et al. 1985). This strongly suggests that the interaction of pp60^{src} with cellular membranes is hydrophobic in character and points out the importance of the association of the protein with membranes in transformation.

3.4 *Sites of Phosphorylation*

p60$^{v\text{-}src}$ contains two major sites of phosphorylation, serine 17 and tyrosine 416, and several minor ones (COLLETT et al. 1979b; SMART et al. 1981; PATSCHINSKY et al. 1982). Serine 17 contains approximately 0.5 moles of phosphate per mole of p60^{src} (SEFTON et al. 1982) and is phosphorylated by the cAMP-dependent protein kinase (COLLETT et al. 1979b). Tyrosine 416 is phosphorylated less extensively containing approximately 0.25 moles of phosphate per mole of polypeptide (SEFTON et al. 1982). It may be subject to autophosphorylation in transformed cells (PURCHIO 1982). The p60^{src} encoded by the Prague strain of RSV contains a second site of tyrosine phosphorylation, most likely tyrosine 205 or 208 (T. PATSCHINSKY and B. SEFTON, unpublished results). Treatment of transformed cells with vanadyl ions reveals the existence of another, normally quite minor, site of tyrosine phosphorylation located somewhere within the amino terminal 160 residues of the protein (COLLETT et al. 1984; BROWN and GORDON 1984).

The question of whether the phosphorylation of p60^{src} regulates its activity has not yet been answered satisfactorily. Deletion of three amino acids, including serine 17, has no effect on the activity of p60^{src} as a protein kinase, when assayed in vitro, or on the ability of the protein to induce the transformation of fibroblasts in culture (CROSS and HANAFUSA 1983). Nevertheless, ROTH et al. (1983) have reported that treatment of RSV-transformed vole cells with agents which should stimulate the cAMP-dependent protein kinase, and hence the phosphorylation of residue 17, increase the protein kinase activity of p60^{src}.

A role for the phosphorylation of tyrosine 416 is clearer. Mutation of this residue to phenylalanine has no effect on protein kinase activity, assayed in vitro, or on the ability of the virus to transform cultured mouse fibroblasts (SNYDER et al. 1983; SNYDER and BISHOP 1984). It does, however, reduce the ability of mouse cells which have been transformed in culture to form tumors in syngeneic mice. This reduced tumorigenicity does not derive from an inability of the injected cells to replicate, but rather from the ability of the mice to reject the tumors. Cells transformed by this mutant p60^{src} are fully tumorigenic in nude mice (SNYDER and BISHOP 1984).

Similarly, a three amino acid deletion encompassing residue 416 has been found to slow the transformation of fibroblasts in culture and to reduce the rate of tumor formation in birds without having a measurable effect on in vitro protein kinase activity of p60^{src} (CROSS and HANAFUSA 1983).

Perhaps most interesting is the recent observation that treatment of cells with vanadyl ions both increases the phosphorylation of p60^{src} on tyrosine and stimulates markedly its protein kinase activity, when assayed in vitro (COLLETT et al. 1984; BROWN and GORDON 1984). Identification of the sites which become phosphorylated is clearly of much interest.

3.5 Substrates of p60^{src}

Chick cells transformed by RSV contain eight cellular polypeptides which are newly phosphorylated on tyrosine. These are vinculin, p81, p50, enolase, p42, p36, lactate dehydrogenase, and phosphoglycerate mutase (SEFTON et al. 1981b; K. GOULD, personal communication; HUNTER and SEFTON 1980; COOPER and HUNTER 1981a; COOPER et al. 1983). Less is known about RSV-transformed mammalian cells. The phosphorylation of vinculin, p36, and phosphoglycerate mutase on tyrosine is observed reproducibly (SEFTON et al. 1981b; COOPER and HUNTER 1981b). Phosphorylated enolase and lactate dehydrogenase are found occasionally (HUNTER and COOPER 1983) and it is likely that p81 will be found also to be phosphorylated on tyrosine in these cells. p42, the ubiquitous substrate of mitogen-activated cellular tyrosine protein kinases is, however, not detectably phosphorylated on tyrosine in any RSV-transformed mammalian cell line examined to date (COOPER and HUNTER 1981b).

3.6 p60$^{c\text{-}src}$

The cellular homologue of p60$^{v\text{-}src}$, p60^{c-src}, is encoded by a large gene which is interrupted by ten intervening sequences (TAKEYA and HANAFUSA 1983). Transduction led to the precise excision of all of the introns. Because the v-*src* gene is flanked on both sides by cellular sequences which are not present in the transcript of the c-*src* gene, it is apparent that RSV arose as the result of recombination between avian leukosis virus DNA and the chromosomal c-*src* locus of the chicken, rather than by recombination between reverse transcripts of the viral genome and the c-*src* mRNA (TAKEYA and HANAFUSA 1983).

p60$^{c\text{-}src}$ is present at a low level in most vertebrate cells (COLLETT et al. 1979a; OPPERMANN et al. 1979). Like p60$^{v\text{-}src}$, it exhibits tyrosine protein kinase activity when assayed in vitro (OPPERMANN et al. 1979; HUNTER and SEFTON 1980) and is itself phosphorylated on tyrosine in vivo (COLLETT et al. 1980). Nevertheless, it is obvious that the expression of p60$^{c\text{-}src}$ at low levels in uninfected cells does not lead to transformation. Until very recently, it was impossible to decide whether p60$^{c\text{-}src}$ was benign simply because its constitutive level of expression was much lower than that of p60$^{v\text{-}src}$ in virally infected cells or because it differed in some fundamental from p60$^{v\text{-}src}$. The molecular cloning of the c-*src* gene has now allowed this question to be answered.

Although encoded entirely by acquired cellular sequences, p60$^{v\text{-}src}$ is not colinear with p60$^{c\text{-}src}$ (Fig. 2). The v-*src* gene contains a deletion which has removed the sequences encoding the 19 carboxy terminal residues of p60$^{c\text{-}src}$ (TAKEYA et al. 1982; SWANSTROM et al. 1983). The carboxy terminus of p60$^{v\text{-}src}$ consists of 12 heterologous residues encoded by sequences normally located 900 nucleotides downstream from the true 3′ end of the c-*src* gene. With this exception, the two proteins are extremely similar in sequence. p60$^{v\text{-}src}$ of Schmidt-Ruppin RSV differs in sequence from p60$^{c\text{-}src}$ at only 8 of 514 positions outside of the dissimilar carboxy terminal region (CZERNILOFSKY et al. 1980, 1983; TAKEYA and HANAFUSA 1983). As a consequence of these differences in sequence, the sites of tyrosine phosphorylation in the two proteins are different. Tyrosine 416 is present in p60$^{c\text{-}src}$ but is not phosphorylated (SMART et al. 1981; KARESS and HANAFUSA 1981). Instead, another, as yet unidentified, tyrosine in the catalytic domain of pp60$^{c\text{-}src}$ is phosphorylated.

These differences in sequence between the viral and cellular proteins are of more than academic interest. Cells expressing high levels of p60$^{c\text{-}src}$, either because they are infected with a virus encoding p60$^{c\text{-}src}$ or because they were transfected with c-*src* DNA, show little evidence of transformation and do not contain noticeably elevated levels of phosphotyrosine in protein (IBA et al. 1984; PARKER et al. 1984; SHALLOWAY et al. 1984; COOPER et al. 1985). Apparently, even high levels of p60$^{c\text{-}src}$ are unable to carry out the phosphorylation of the substrates critical to cellular transformation.

Replacement of the carboxy terminus of p60$^{c\text{-}src}$ with that of p60$^{v\text{-}src}$ greatly increases the transforming activity of p60$^{c\text{-}src}$ (IBA et al. 1984; SHALLOWAY et al. 1984). This suggests that the deletion of the c-*src* carboxy terminus plays an important role in the activation of the gene. There is also, however, suggestive evidence that one or more of the changes in sequence outside of the carboxy terminus also increase the transforming activity of p60^{src} (IBA et al. 1984; PARSONS et al. 1984). These have not yet been localized precisely.

p60$^{c\text{-}src}$ may, however, have oncogenic potential. Cells transformed by the murine DNA tumor virus polyoma virus contain a subpopulation of p60$^{c\text{-}src}$ which is bound to the presumptive transforming protein of polyoma virus, the medium T antigen (COURTNEIDGE and SMITH 1983). This population of p60$^{c\text{-}src}$ is several times more active as a tyrosine protein kinase, when assayed in vitro (BOLEN et al. 1984). The increased tyrosine protein kinase activity of this form of p60$^{c\text{-}src}$ could play a role in the process by which polyoma virus induces transformation. Evidence of increased tyrosine protein kinase activity in polyoma virus transformed cells has not, however, yet been obtained (SEFTON et al. 1980; COOPER and HUNTER 1981 b).

The c-*src* gene is recognizable and expressed in organisms as primitive as *Drosophila* (HOFFMAN et al. 1983; SIMON et al. 1983). Here, the expression of the gene is regulated in a pronounced fashion during development (LEV et al. 1984). p60$^{c\text{-}src}$ is present in most avian and mammalian cells (COLLETT et al. 1979a; OPPERMANN et al. 1979). It is unusually abundant in the brain and in neural retinal cells (COTTON and BRUGGE 1983; SORGE et al. 1984). There is no indication to date that the levels of the protein are regulated during vertebrate development or as a result of physiological challenge.

4 *abl*

The *abl* oncogene is so named because it was first identified in Abelson murine leukemia virus, a murine retrovirus which induces a B cell disease in mice and can transform both fibroblastic and hematopoetic cells in culture (SCHER and SIEGLER 1975; ROSENBERG et al. 1975). It was isolated originally from a lymphosarcoma which arose in a mouse injected with Moloney leukemia virus (ABELSON and RABSTEIN 1970). Abelson virus is unique among the viruses which encode tyrosine protein kinases in that it induces B cell disease and a thorough discussion of it can be found in the article by WITTE (1983) in Vol. 103 of this series. A feline sarcoma virus which also carries the *abl* oncogene has recently been identified (Table 3; BESMER et al. 1983).

4.1 Structure of the Abelson Virus Transforming Protein

The original isolate of Abelson virus encodes a single chimeric polypeptide termed either P160^{abl} or P160$^{gag\text{-}abl}$ (ROSENBERG and WITTE 1980). The amino terminal 236 residues of P160^{abl} are encoded by a fragment of the Moloney leukemia virus *gag* gene (REDDY et al. 1983). The carboxy terminal 1008 residues are encoded by a large portion of the c-*abl* gene of the mouse (Fig. 3; SHIELDS et al. 1979; REDDY et al. 1983). Variants of Abelson virus encoding truncated forms of P160^{abl} arise with a surprising frequency.

P160^{abl} is comprised of three domains. The amino terminal 236 amino acids are identical in sequence to the amino terminal half of Pr65^{gag}, the precursor

Table 3. Viruses which encode tyrosine protein kinases

	Transforming protein	Animal of origin
src		
Rous sarcoma virus	p60^{src}	Chicken
abl		
Abelson murine leukemia virus	P160$^{gag\text{-}abl}$	Mouse
Hardy-Zuckerman-2 feline sarcoma virus	P98$^{gag\text{-}abl}$	Cat
*erb*B		
Avian erythroblastosis virus	gp74^{erbB}	Chicken
fps		
Fujinami sarcoma virus	P140$^{gag\text{-}fps}$	Chicken
PRCII virus	P105$^{gag\text{-}fps}$	Chicken
PRCIV virus	P170$^{gag\text{-}fps}$	Chicken
UR1 virus	P150$^{gag\text{-}fps}$	Chicken
16L virus	P142$^{gag\text{-}fps}$	Chicken
Snyder-Theilen feline sarcoma virus	P85$^{gag\text{-}fps(fes)}$	Cat
Gardner-Arnstein feline sarcoma virus	P95/105$^{gag\text{-}fps(fes)}$	Cat
yes		
Y73 virus	P90$^{gag\text{-}yes}$	Chicken
Esh sarcoma virus	P80$^{gag\text{-}yes}$	Chicken
fgr		
Gardner-Rasheed feline sarcoma virus	P70$^{gag\text{-}fgr}$	Cat
ros		
UR2 virus	P68$^{gag\text{-}ros}$	Chicken
fms		
McDonough feline sarcoma virus	gp140^{fms}	Cat

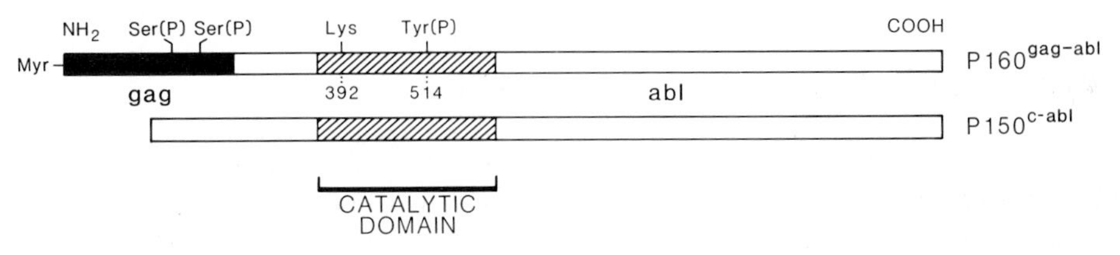

Fig. 3. Comparison of P160$^{v\text{-}abl}$ and P150$^{c\text{-}abl}$. The amino terminal portion of the v-*abl* protein, which is identical to the amino terminal portion of the *gag* gene of the parental Moloney leukemia virus, is shown in *black*. The regions encoded by *abl* sequences are shown in *white*. The catalytic domain of the two proteins is indicated by *crosshatching*. Indicated also is the myristyl moiety present in the viral protein, the two phosphorylated serines present in the *gag*-derived region of the viral protein, lysine 392 (located in the predicted ATP-binding site), and tyrosine 514 (phosphorylated in the viral protein, but not in the cellular protein)

to the internal structural proteins of the murine C-type retroviruses. In leukemia virus virions, the two proteins encoded by this region of the *gag* gene serve to condense the viral genomic RNA and promote the interaction of the viral core particle with the virion envelope. What function this fragment of $Pr65^{gag}$ performs in $P160^{abl}$ is unknown. Most of this region is not essential for the transformation of fibroblasts (PRYWES et al. 1983).

Residues 356 to 607 comprise the catalytic domain. This is apparent both from their marked homology with the catalytic domain of $p60^{src}$, which includes a lysine in the same position as the lysine in $p60^{src}$ which has been shown to participate in binding ATP (KAMPS et al. 1984), and from the fact that deletions which extend into this region ablate protein kinase activity (PRYWES et al. 1983). Additionally, in *E. coli*, DNA encoding only this region of the protein can direct the synthesis of a polypeptide fragment which possesses tyrosine protein kinase activity (WANG and BALTIMORE 1985).

The carboxy terminal 637 residues comprise a third domain of unknown function. Deletions in this region do not affect either tyrosine protein kinase activity or the ability of the virus to transform fibroblasts, (ROSENBERG et al. 1980; PRYWES et al. 1983). They do, however, render the Abelson protein unstable in lymphoid cells (PRYWES et al. 1985).

4.2 Subcellular Location of P160^abl

Immunofluorescence with monoclonal anti-*gag* antibodies detects $P160^{abl}$ in the cytoplasm of transformed fibroblasts where it is associated with the plasma membrane and concentrated in adhesion plaques (ROHRSCHNEIDER and NAJITA 1984).

$Pr65^{gag}$, the *gag* gene precursor of Moloney leukemia virus, contains myristic acid at this amino terminus (SCHULTZ and OROSZLAN 1983). Since the amino terminal domains of $P160^{abl}$ and $Pr65^{gag}$ are identical, $P160^{abl}$, like $p60^{src}$, is myristylated (SCHULTZ and OROSZLAN 1984). It is not yet known whether this plays an important role in the transforming activity of the protein or merely reflects the ancestry of $P160^{abl}$. This is a question worth pursuing.

4.3 Sites of Phosphorylation

$P160^{abl}$ is multiply phosphorylated. It contains four sites of serine phosphorylation, two sites of threonine phosphorylation, and two sites of tyrosine phosphorylation (SEFTON et al. 1981a). Two of the phosphorylated serines are located in the region of the amino terminal *gag* domain which is homologous to the structural phosphoprotein of Moloney leukemia virus, $p12^{gag}$ (VAN DE VEN et al. 1980a; PATSCHINSKY and SEFTON 1981). One of the sites of tyrosine phosphorylation, tyrosine 514 (PATSCHINSKY et al. 1982) is located within the catalytic domain of the protein at a site homologous to that of the single phosphorylated tyrosine in $p60^{src}$.

The identities of the protein kinases phosphorylating $P160^{abl}$ are uncertain. $P160^{abl}$ undergoes extensive autophosphorylation in vitro; one of the sites phosphorylated is tyrosine 514 (REYNOLDS et al. 1982). This site may also be subject to autophosphorylation in vivo. The question of whether the phosphorylation of $P160^{abl}$ regulates its activity has not yet been addressed.

4.4 *Polypeptide Substrates*

Vinculin, p36, and phosphoglycerate mutase are always found to be phosphorylated on tyrosine in fibroblasts transformed by Abelson virus (SEFTON et al. 1981b; COOPER and HUNTER 1981b). On occasion, the phosphorylation of p81, enolase, and lactate dehydrogenase on tyrosine has been detected (HUNTER and COOPER 1983). The phosphorylation of cellular proteins in transformed lymphocytes is much less extensive; only p36 and phosphoglycerate mutase have been found to contain elevated levels of phosphotyrosine (SEFTON et al. 1983). Phosphorylated p36 is not present in all Abelson-transformed lymphoid cell lines. A surprising number of both normal and transformed lymphoid cell lines contain no p36 polypeptide (SEFTON et al. 1983). The phosphorylation of p42 has not been detected in either transformed fibroblasts or transformed lymphoid cells.

4.5 *c-abl*

The cellular homologue of v-*abl* is encoded by a gene encompassing more than 30 kilobases of cellular DNA in the mouse (WANG et al. 1984). It is interrupted by more than nine intervening sequences, each of which was excised precisely during the genesis of the virus. The murine c-*abl* locus encodes a protein of 150000 daltons, termed NCP150 or P150 (WITTE et al. 1979a). The human locus encodes a slightly smaller protein, termed P145 (KONOPKA et al. 1984). The viral and cellular *abl* proteins appear to be coterminal at their carboxy termini (Fig. 3). A small portion of the amino terminus of the c-*abl* protein was lost during the genesis of $P160^{v\text{-}abl}$ and was replaced with viral *gag* sequences (WANG et al. 1984). Both P145 or P150 possess tyrosine protein kinase activity in vitro (O. WITTE, personal communication). Unlike $p60^{c\text{-}src}$, neither is phosphorylated on tyrosine in vivo (PONTICELLI et al. 1982; KONOPKA et al. 1984).

The chromosomal c-*abl* locus has been found to have undergone rearrangement in a number of human cell lines derived from patients with chronic myelogenous leukemia (COLLINS and GROUDINE 1983; HEISTERKAMP et al. 1983). Such cell lines encode two c-*abl* gene products; P145 and a larger protein termed P210 (KONOPKA et al. 1984). Unlike NCP145, P210 is phosphorylated on tyrosine in cultured cells.

Apparently, the rearrangement of the c-*abl* locus in these cells leads to replacement of the normal amino terminal domain of P145, producing P210.

The tyrosine protein kinase activity of P210 is more stable in vitro than that of NCP145. This could reflect an altered activity of the protein in vivo.

Organisms as primitive as *Drosophila melanogaster* possess a c-*abl* gene (HOFFMAN et al. 1983). Here, its expression is regulated during embryogenesis (LEV et al. 1984). The gene is expressed in the developing mouse embryo and most adult tissues (MÜLLER et al. 1982; WANG and BALTIMORE 1983); c-*abl* transcripts are particularly abundant in testes and fibroblasts.

5 *erb*B

The *erb*B oncogene is one of two completely unrelated cellular genetic sequences present in the genome of the ES4 strain of avian erythroblastosis virus. This oncogene is apparently sufficient for the induction of erythroblastosis in the chicken, since it is the only oncogene present in the genome of a second independently isolated avian erythroblastosis virus (YAMAMOTO et al. 1983a). *erb*B is of particular interest because it is derived from the gene for the chicken EGF receptor (DOWNWARD et al. 1984a).

5.1 Structure of the erbB Protein

Cells transformed by avian erythroblastosis virus contain multiple forms of the *erb*B gene product (HAYMAN et al. 1983; PRIVALSKY et al. 1983). This heterogeneity reflects the fact that the *erb*B product is a glycoprotein whose oligosaccharides rarely undergo complete naturation. The unglycosylated protein is 61000 daltons in size. Highly glycosylated forms with apparent molecular weights as great as 82000 daltons have been described (DECKER 1985). The form of the protein exposed on the cell surface is usually termed gp74^{erbB} (HAYMAN and BEUG 1984).

gp74^{erbB} is a transmembrane cell surface glycoprotein which is composed of three very different domains. The amino terminal 65 residues form the extracellular domain (YAMAMOTO et al. 1983b). This region contains three potential sites for the attachment of *N*-glycosidically linked oligosaccharides to asparagine, residues 24, 44, and 59. All of these appear to be occupied. Residues 66 to 88 most probably comprise the transmembrane segment. The remainder of the protein, residues 89 to 604, are almost certainly present on the cytoplasmic face of the plasma membrane. Comparison of the deduced sequence of gp74^{erbB} with that of p60^{src} suggests that residues 135 to 386 form the catalytic domain of the protein. By analogy with p60^{src}, lysine 166 is predicted to comprise part of the ATP-binding site (KAMPS et al. 1984).

Neither of the two predicted amino termini of the v-*erb*B protein bears much resemblance to the signal peptides of most proteins destined to be transported to the cell surface (YAMAMOTO et al. 1983b). How translocation of the amino terminal portion of gp74^{erbB} across the membrane of the endoplasmic reticulum occurs is not understood. The five carboxy terminal residues of gp74^{erbB} are not encoded by the acquired *erb*B element but rather by a fragment of the *env* gene of an avian leukosis virus.

5.2 Subcellular Location of the Polypeptide Products of erbB

Both cell fractionation and immunofluorescence demonstrate that gp74^{erbB} is a cell surface glycoprotein (HAYMAN and BEUG 1984). Nevertheless, the bulk of the *erb*B proteins are both smaller than gp74^{erbB} and are found associated with intracellular membranes (HAYMAN et al. 1983; PRIVALSKY et al. 1983). Some of these proteins represent precursors to gp74^{erbB} which are undergoing transport to the cell surface. Both maturation and glycosylation of the *erb*B proteins appear, however, to be very inefficient. The fraction of *erb*B proteins which reach the cell surface is small. Most are degraded.

5.3 Sites of Phosphorylation

gp74^{erbB} is poorly phosphorylated in vivo and contains little, if any, phosphotyrosine (HAYMAN et al. 1983). Under some circumstances, it will undergo autophosphorylation on tyrosine in vitro (DECKER 1985; GILMORE et al. 1985; KRIS et al. 1985). The identity of the tyrosine residue which becomes phosphorylated is currently unknown. Decker has reported that treatment of AEV-transformed cells with TPA induces both phosphorylation of gp74^{erbB} on threonine and inhibits the growth of the transformed cells (DECKER 1985). TPA also induces the phosphorylation of the c-*erb*B product, the EGF receptor, at threonine 654 (HUNTER et al. 1984), and inhibits its activity (COCHET et al. 1984; FRIEDMAN et al. 1984; IWASHITA and FOX 1984). By analogy, it is possible that TPA induces the phosphorylation of gp74^{erbB} on threonine 99 and that this inhibits the viral protein.

5.4 Polypeptide Substrates

Initial attempts to demonstrate tyrosine protein kinase activity associated with gp74^{erbB} isolated by immunoprecipitation yielded negative results (HAYMAN et al. 1983). Recent efforts have, however, detected an activity catalyzing the phosphorylation of gp74^{erbB} on tyrosine in immunoprecipitates containing the protein (DECKER 1985; KRIS et al. 1985). The explanation for these contradictory findings is not obvious. It is conceivable that the antibodies used in the first experiments inhibited autophosphorylation.

Evidence that was consistent with one of the gene products of AEV ES4 possessing tyrosine protein kinase activity was in fact obtained several years ago. Phosphorylated p36 was detected by two-dimensional gel electrophoresis of the phosphoproteins of AEV-transformed fibroblasts (RADKE and MARTIN 1979) and analysis of total cellular phospho-amino acids in protein revealed a modest 50% increase in the abundance of phosphotyrosine in protein in

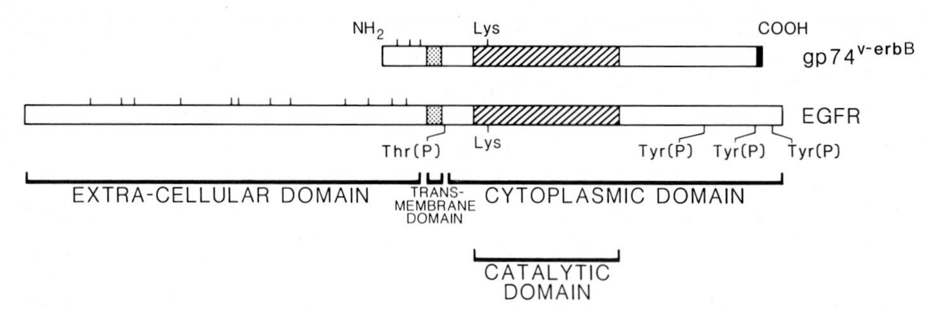

Fig. 4. Comparison of gp74$^{v\text{-}erbB}$ and the EGF receptor, the products of the v-*erb*B and c-*erb*B genes. The predicted membrane-spanning domains of the two proteins is indicated by *stippling*, the predicted catalytic domains are indicated by *crosshatching*, and the carboxy terminus of the viral protein (encoded by a fragment of the viral *env* gene) is shown in *black*. The potential glycosylation sites in the extracellular domains of the two proteins are shown with *short vertical lines*. Also indicated are threonine 654 (phosphorylated in the EGF receptor by protein kinase C), tyrosines 1068, 1148, and 1173 (phosphorylated when the receptor undergoes autophosphorylation in vitro), and lysine 722 (predicted to lie in the ATP-binding site)

these cells (SEFTON et al. 1980). Recent careful analysis of protein phosphorylation in AEV-transformed fibroblasts using alkali-treated two-dimensional gels shows that these cells contain both significantly elevated levels of phosphorylated p42, some phosphorylated enolase, and a modest level of phosphorylated p36 (GILMORE et al. 1985). It is notable that treatment of uninfected chick cells with epidermal growth factor stimulates the phosphorylation of p42, but not of p36 and enolase (NAKAMURA et al. 1983; GILMORE and MARTIN 1983; COOPER et al. 1984).

5.5 *The Polypeptide Product of the c-erb*B *Gene*

The v-*erb*B gene is a drastically truncated form of the gene for the EGF receptor (Fig. 4; DOWNWARD et al. 1984a). Missing from gp74^{erbB} are 556 amino terminal residues which comprise most of the extracellular EGF-binding domain of the receptor and 32 residues from the carboxy terminus of the receptor (ULLRICH et al. 1984).

gp74^{erbB} and the EGF receptor are very different proteins. Both the tyrosine protein kinase activity and the phosphorylation of the receptor at tyrosine 1173 are stimulated markedly by the binding of EGF (USHIRO and COHEN 1980; DOWNWARD et al. 1984b). In contrast, gp74^{erbB} contains only a stub of the extracellular EGF-binding domain of the receptor and lacks the carboxy terminal portion of the receptor which contains tyrosine 1173. It is unlikely, therefore, that gp74^{erbB} is subject to the same regulation as is the EGF receptor. It could be locked into an active conformation.

The EGF receptor is present in many tissues, not just those of epithelial origin (FABRICANT et al. 1977). Like the c-*src*, c-*abl*, and c-*ras* oncogenes, the gene for the EGF receptor has been found to be present and to be expressed in primitive organisms such as *Drosophila* (LIVNEH et al. 1985).

6 *fps*

The *fps* oncogene has appeared in the genomes of at least seven independently isolated acutely transforming retroviruses. This family of viruses includes the avian sarcoma viruses, Fujinami sarcoma virus (FSV), PRCII virus, PRCIV viruses, 16L virus, and UR1 virus, and the Gardner-Arnstein (Ga-FeSV) and Snyder-Theilen (ST-FeSV) feline sarcoma viruses (LEE et al. 1980; HANAFUSA et al. 1980; PAWSON et al. 1980; SHIBUYA et al. 1980; BARBACID et al. 1980; NEIL et al. 1981a; BREITMAN et al. 1981; NEEL et al. 1982; WANG et al. 1982a). The oncogene transduced by this family of avian viruses from the chicken genome is called *fps* and that acquired by the two feline viruses from the cat genome, *fes*. It has recently been proven that these two genes are true homologues (GROFFEN et al. 1983). The single gene carried by all of these viruses will be referred to here as *fps*. None of these viruses carries the complete c-*fps* gene (see below). The portion of the c-*fps* gene in the genome of Fujinami sarcoma virus is the largest and the majority of the discussion here will therefore focus on the transforming protein of this virus.

6.1 Structure of the Transforming Protein of FSV

Like all of the *fps*-containing viral polypeptides, the transforming protein of FSV, P140^{fps}, is a hybrid consisting of an amino terminal domain encoded by the *gag* gene of a leukosis virus and a carboxy terminal domain encoded by the transduced *fps* sequences. It is, therefore, sometimes referred to as P140$^{gag\text{-}fps}$. The structures of this family of transforming proteins, have fundamental similarity to that of P160^{abl}. P140^{fps} consists of 308 residues identical in sequence to the amino terminal domain of the Pr76^{gag} protein of the avain leukosis viruses and 874 residues encoded by the acquired c-*fps* sequences (SHIBUYA and HANAFUSA 1982). The protein is terminated within the acquired cellular sequences. Since the *fps* sequences in the genome of FSV encode approximately 95000 daltons protein and the c-*fps* product is 98000 daltons in size (MATHEY-PREVOT et al. 1982), it is likely that almost all of the c-*fps* gene is represented in the genome of FSV.

Comparison of the sequence of P140^{fps} with that of p60^{src} reveals that residues 901 to 1168, which are located at the carboxy terminus of the protein, comprise the catalytic domain. The exact function of the other portions of this large transforming protein are not well understood. PRCII virus appears to be a natural deletion mutant of FSV. Its transforming protein, P105^{fps}, lacks residues 413 to 753 of P140^{fps} (DUESBERG et al. 1983; CARLBERG et al. 1984). Despite this large deletion, P105^{fps} exhibits tyrosine protein kinase activity, both in vitro and in vivo, and transforms cultured fibroblasts (NEIL et al. 1981a), albeit less dramatically than does FSV (GUYDEN and MARTIN 1982). PRCII virus is, however, measurably less tumorigenic than FSV in birds (DUESBERG et al. 1983). The genomes of both GA-FeSV and ST-FeSV also lack the portion of the *fps* gene which encodes residues 529–707 of P140^{fps} (HAMPE et al. 1982). This region of P140^{fps}, therefore, apparently can facilitate, but is certainly not essential for, either protein kinase activity or cellular transformation.

The amino terminal *gag* gene-derived portion of P140^{fps} is also apparently dispensible. Deletion of all of the *gag* gene sequences from the genome of FSV yields a virus which retains the ability to transform cells in culture and to induce tumors in birds (FOSTER and HANAFUSA 1983). Cells transformed by this virus, however, undergo a less dramatic morphological transformation than do cells transformed by wild-type FSV.

6.2 Subcellular Location of P140^{fps}

The subcellular location of P140^{fps} differs from that of the other well-studied viral tyrosine protein kinases. Immunofluorescence with anti-*fps* antisera shows that the protein is present diffusely in the cytoplasm of transformed cells and is not obviously associated with the cellular plasma membrane, adhesion plaques or cell:cell junctions (FELDMAN et al. 1983; WOOLFORD and BEEMON 1984; MOSS et al. 1984). Furthermore, although the protein can be recovered in fractions containing plasma membranes, if cell fractionation is carried out with buffers of low ionic strength, P140^{fps} behaves as a soluble protein at higher salt concentrations (FELDMAN et al. 1983; MOSS et al. 1984). If P140^{fps} is bound to the plasma membrane, it is not held there by hydrophobic forces. P140^{fps}, therefore, bears less resemblance to a typical membrane protein than do p60^{src} and gp74^{erbB}. P140^{fps} is not myristylated (B. SEFTON, unpublished results). In contrast, because they possess amino termini encoded by fragments of the 5′ end of the *gag* gene of feline leukemia viruses, the transforming proteins of both GA-FeSV and ST-FeSV are myristylated (SCHULZ and OROSZLAN 1984). Whether this causes them to bind more strongly to cellular membranes than P140^{fps} is currently unknown.

6.3 Sites of Phosphorylation

P140^{fps} contains a minor site of tyrosine phosphorylation and one phosphorylated serine in the *gag*-encoded domain and two phosphorylated tyrosines and one phosphorylated serine in the *fps*-encoded domain (WEINMASTER et al. 1983). The only identified site of phosphorylation, tyrosine 1073, is located within the catalytic domain at a site homologous to that of the single phosphorylated tyrosine found in p60^{src}. The phosphorylation of this site in P140^{fps} plays a more important role than does the phosphorylation of the homologous site in p60^{src}. Mutation of this residue to phenylalanine creates a strain of FSV which, when transfected into rodent fibroblasts in culture, induces transformation only after a pronounced lag (WEINMASTER et al. 1984). In addition, unlike the similar mutation in p60^{src}, conversion of this residue to phenylalanine reduces the ability of P140^{fps} to phosphorylate exogenous substrates in vitro five-fold. It would appear that a phosphorylated residue at this site increases the inherent protein kinase activity of P140^{fps} and that this, in turn, increases the oncogenicity of the virus.

6.4 *Polypeptide Substrates*

The phosphorylation of cellular proteins induced by viruses carrying the *fps* gene has been studied carefully in both avian and rodent fibroblasts. The phosphorylation of p36, p81, enolase, p42, phosphoglycerate mutase, and lactate dehydrogenase has been observed (COOPER and HUNTER 1981 b; HUNTER and COOPER (1983). In contrast, vinculin is poorly phosphorylated on tyrosine (SEFTON et al. 1981 b). Apparently the *fps* gene products interact poorly with this cellular substrate. The lack of phosphorylation of vinculin could contribute to the fact that avian fibroblasts transformed by viruses carrying this gene never become as fully rounded as do cells transformed by RSV.

6.5 *The c-fps Product*

The c-*fps* gene encodes a product of approximately 98000 daltons, termed NCP98 or NCP98$^{c\text{-}fps}$ (BARBACID et al. 1980 a; MATHEY-PREVOT et al. 1982). The viral and cellular *fps* proteins are coterminal at their carboxy termini (SHIBUYA and HANAFUSA 1982) and differ principally in that a small portion of the amino terminus of the cellular protein is replaced by *gag*-derived sequences in the viral protein. This is very similar to what is seen when the viral and cellular *abl* proteins are compared. Like the viral *fps*-containing proteins, NCP98 possesses tyrosine protein kinase activity when assayed in vitro. Efforts to determine whether the protein is itself phosphorylated on tyrosine in vivo have not yielded conclusive results.

NCP98$^{c\text{-}fps}$ is found predominantly in the cytosol when chicken myeloblasts are subjected to traditional cell fractionation (YOUNG and MARTIN 1984). Its solubility shows much less dependence on the ionic strength of the extraction buffer than does the solubility of P140$^{v\text{-}fps}$.

c-*fps* is expressed most actively in the bone marrow and least actively in muscle and heart (SHIBUYA et al. 1982). Expression of the feline c-*fps* gene in fibroblasts has, however, been reported (BARBACID et al. 1980 a). It is likely that the *Drosophila* genome contains a c-*fps* homologue, but this gene has not yet been characterized (HOFFMAN et al. 1983).

7 *yes*

The *yes* oncogene is so named because it was found in the genomes of two avian sarcoma viruses, Yamaguchi 73 virus (Y73 virus) and Esh sarcoma virus (ESV) (YOSHIDA et al. 1980; GHYSDAEL et al. 1981). Like RSV, these viruses arose in a chicken as a result of the recombination between the genome of an avian leukosis virus and chicken genetic sequences.

7.1 *Structure of the Viral Transforming Protein*

Y73 virus encodes a single hybrid polypeptide termed either P90^{yes} or P90$^{gag\text{-}yes}$. The structure of P90^{yes} resembles somewhat that of P160$^{gag\text{-}abl}$ and P140$^{gag\text{-}fps}$.

This protein consists of 220 amino terminal amino acids derived from the 5′ end of the *gag* gene of its viral ancestor, 585 from the chicken c-*yes* gene, and seven carboxy terminal residues which are encoded in a normally unused reading frame by a fragment of the viral *env* gene (KITAMURA et al. 1982). By analogy with $p60^{src}$, the catalytic domain of the protein appears to be comprised of residues 500–800 which are located at the carboxy terminus of the protein.

The acquired *yes* sequences exhibit extraordinary homology with *src*. Within one portion of the catalytic domain, $P90^{yes}$ and $p60^{src}$ have 93% sequence identity. The homology is also striking in more amino terminal regions of the two proteins which do not form part of the catalytic domain. Residues 365–554 of $P90^{yes}$ have 82% sequence identity with residues 81–250 of $p60^{src}$. No homology, however, is apparent between the membrane-binding domain of $p60^{src}$ and the amino terminal portion of $P90^{yes}$.

7.2 Subcellular Location of $P90^{yes}$

Immunofluorescence with antipeptide antibodies cross-reactive with $P90^{yes}$ indicate that the protein is distributed in transformed chick cells in much the same manner as $p60^{src}$ (GENTRY and ROHRSCHNEIDER 1984). $P90^{yes}$ apparently resides exclusively in the cytoplasm, where it is concentrated at the cytoplasmic face of the plasma membrane in adhesion plaques and at cell:cell junctions. Unlike $p60^{src}$, $P90^{yes}$ contains no lipid (B. SEFTON, unpublished results) and the manner with which it binds to the plasma membrane is not understood.

7.3 Sites of Phosphorylation

$P90^{yes}$ contains two sites of tyrosine phosphorylation and two sites of serine phosphorylation (PATSCHINSKY and SEFTON 1981). One phosphorylated serine is in the *gag*-derived domain, presumably within the region which corresponds to $p19^{gag}$. Tyrosine 700, which is located within the catalytic domain in a position homologous to the single phosphorylated tyrosine in $p60^{src}$, is the only identified site of phosphorylation (PATSCHINSKY et al. 1982; NEIL et al. 1981b; KITAMURA et al. 1982). The identities of the kinases phosphorylating $P90^{yes}$ in vivo are unknown. Tyrosine 700 is phosphorylated in vitro and may therefore be phosphorylated by autophosphorylation in vivo.

7.4 Polypeptide Substrates

Transformation of chick cells by Y73 virus induces the phosphorylation on tyrosine of vinculin, enolase, p42, and p36 (SEFTON et al. 1981b; COOPER and HUNTER 1981b). Whether p81, phosphoglycerate mutase, and lactate dehydrogenase also undergo increased phosphorylation on tyrosine has not been determined.

7.5 *The Polypeptide Product of c-yes*

The product of the c-*yes* gene has not been identified. Because the fragment of the *yes* gene in the genome of Y73 virus encodes 585 amino acids, the c-*yes* product will almost certainly prove to be larger than 65000 daltons. c-*yes* is known to be expressed in most avian tissues, at a level significantly higher than that of c-*src* (SHIBUYA et al. 1982). c-*yes* transcripts are unusually abundant in the kidney. It is not yet known whether the gene is present in the genome of invertebrates.

8 *fgr*

The *fgr* oncogene is carried by the Gardner-Rasheed strain of feline sarcoma virus (GR-FeSV; NAHARRO et al. 1984). The transforming protein of this virus is unusual among those discussed here in that it contains segments encoded by two unrelated cellular genes. It is termed either $P70^{fgr}$, $P70^{gag\text{-}fgr}$, or $P70^{gag\text{-}actin\text{-}fgr}$. This 70000 dalton protein (NAHARRO et al. 1983) consists of 118 amino terminal amino acids identical to the amino terminal domain of the *gag* gene precursor of an unidentified feline leukemia virus, 151 residues encoded by a fragment of the 5′ end of a gene for a feline nonmuscle actin, 389 residues encoded by a fragment of the feline *fgr* gene, and five amino acids encoded by a remnant of the *env* gene of the parental leukemia virus (NAHARRO et al. 1984). In the transduced region, the actin gene apparently includes a portion of the 5′ untranslated region of the actin mRNA which here is expressed as protein. The natural amino terminus of the nonmuscle actin, therefore, resides within the body of $P70^{fgr}$. The sequences to which the name *fgr* refers are not agreed upon. This name is sometimes used to describe the totality of acquired cellular sequences in the viral genome. The term is used here to designate the portion derived from a cellular gene for a tyrosine protein kinase. The expression of the c-*fgr* locus has not been examined at all.

By comparison with the sequence of $p60^{src}$, it is apparent that residues 400–557 form the catalytic domain of $P70^{fgr}$. As with many viral tyrosine protein kinases, the catalytic domain is located at the carboxy terminus of the protein. The function and importance of the other regions of this novel protein have not yet been assessed. *fgr*, which is feline in origin, has remarkable homology with the avian *yes* gene. Despite this species difference, $P90^{yes}$ and $P70^{fgr}$ have 80% sequence identity in certain regions.

8.1 *Subcellular Location of* $P70^{fgr}$

$P70^{fgr}$ has been detected in the cytoplasm of fibroblasts transformed by GR-FeSV by immunofluorescent microscopy using antipeptide antibodies which cross-react with the protein (R. MANGER and L. ROHRSCHNEIDER, personal communication). Subcellular fractionation suggests that some of the protein is particulate, and some soluble. Association with adhesion plaques has not been observed. Because its amino terminus is identical to that of the *gag* gene precursor of the feline leukemia viruses, P70^{fgr} contains myrtistic acid linked to its amino terminal residue (J. BUSS and B. SEFTON, unpublished results). Whether this moiety contributes to the properties of this transforming protein is not yet known.

8.2 Sites of Phosphorylation

P70^{fgr} contains three sites of tyrosine phosphorylation and two sites of serine phosphorylation (T. PATSCHINSKY and B. SEFTON, unpublished results). It is likely that tyrosine 553, which is located in an analagous site to that of the single phosphorylated tyrosine in p60^{src}, is one of the phosphorylated residues. This site is phosphorylated when P70^{fgr} undergoes autophosphorylation in vitro. The kinases which phosphorylate the other sites have not been identified.

8.3 Polypeptide Substrates

The phosphorylation of cellular proteins has been studied only with one line of GR-FeSV-transformed Fisher rat embryo cells. Phosphorylated enolase and phosphorylated p36 were readily detected in these cells (J. COOPER and B. SEFTON, unpublished results). The increased phosphorylation of other cellular proteins on tyrosine was not apparent.

9 *ros*

The *ros* oncogene is found in the genome of the avian sarcoma virus, UR2 virus (WANG et al. 1982b). No other virus carrying this gene has been described. The genome of UR2 virus encodes only a single polypeptide, P68^{ros} (FELDMAN et al. 1982). Like P160^{abl}, P90^{yes}, and P70^{fgr}, P58^{ros} is a hybrid protein comprised of an amino terminal domain identical to the 150 amino terminal residues of the *gag* gene precursor of an avian leukosis virus and 402 carboxy terminal residues encoded by the acquired cellular oncogene. Comparison of the deduced sequence of P68^{ros} with that of p60^{src} suggests that residues 244 to 502 comprise the catalytic domain.

There are a number of indications that P68^{ros} may be a fragment of an as yet unidentified cell surface receptor. It has noticeably more homology with the insulin receptor that it does to other tyrosine protein kinases (ULLRICH et al. 1985). Additionally, residues 158 to 185 of P68^{ros} are all hydrophobic in character and could therefore function as a membrane-spanning domain.

9.1 Subcellular Location of P68^{ros}

Immunofluorescence microscopy using anti-*gag* antiserum shows that P68^{ros} is present in the cytoplasm of infected chicken cells (NOTTER and BALDUZZI 1984). Pronounced concentration of the protein in adhesion plaques or cell:cell junctions has not been noted. Whether the protein associates with cellular membranes is not known. The presence of a potential membrane-spanning domain suggests that it may. $P68^{ros}$ is not myristylated (J. BUSS and B. SEFTON, unpublished results).

9.2 *Sites of Phosphorylation*

$P68^{ros}$ is reported to be phosphorylated in vivo (FELDMAN et al. 1982). Efforts to detect phosphotyrosine in the protein have, however, been unsuccessful (T. PATSCHINSKY and B. SEFTON, unpublished results). The immunoprecipitated protein undergoes autophosphorylation in vitro (FELDMAN et al. 1982), but the identity of the phosphorylated site is not known with certainty.

9.3 *Polypeptide Substrates*

Two-dimensional gel electrophoresis of total cellular phosphoproteins and immunoprecipitation of vinculin have so far failed to detect any specific cellular proteins which contain an increased level of phosphotyrosine in UR2 virus-transformed cells (ANTLER et al. 1985; J. COOPER and B. SEFTON, unpublished results). This is consistent with the observation that UR2-transformed chicken and rat cells contain only slightly more phosphotyrosine in total cellular protein than do normal cells (T. PATSCHINSKY and B. SEFTON, unpublished results). In light of this apparently restricted tyrosine protein kinase activity, it is worth noting that $P68^{ros}$ is reported to possess phosphatidylinositol kinase activity (MACARA et al. 1984). The formal possibility exists, therefore, that it is the phosphorylation of phosphatidylinositol rather than the phosphorylation of polypeptides which is important in cellular transformation by UR2 virus.

Transformation of chick cells by UR2 virus is a subtle process. The cells become slightly elongated, but the pronounced effects induced by other viruses encoding tyrosine protein kinases are not observed (BALDUZZI et al. 1981; NOTTER and BALDUZZI 1984). The virus may have a greater effect in mammalian cells. A number of rat cell lines transformed by UR2 virus have been isolated. Unlike UR2 virus-infected chick cells, they are highly rounded and poorly adherent (T. PATSCHINSKY and B. SEFTON, unpublished results).

In that it contains little phosphotyrosine itself and induces little detectable increase in the phosphorylation of cellular proteins on tyrosine, $P68^{ros}$ resembles the medium T antigen of polyoma virus. Unlike this polyoma T antigen however, the tyrosine protein kinase activity associated with $P68^{ros}$ cosediments with monomeric $P68^{ros}$ (R. VESCIO and B. SEFTON, unpublished results). The activity, therefore, appears to be intrinsic to $P68^{ros}$, rather than to arise from some bound cellular tyrosine protein kinase.

9.4 *c-ros*

The c-*ros* gene is expressed in a number of tissues of the chicken (SHIBUYA et al. 1982). Whether more primitive organisms contain a homologous gene has not been investigated. The polypeptide product of the c-*ros* gene has not yet been identified.

10 *fms*

The *fms* oncogene, which is found in the genome of McDonough feline sarcoma virus, encodes a transforming protein which has traditionally not been included in the list of tyrosine protein kinases. This omission is almost certainly an error. Nucleotide sequence analysis shows that the *fms* product has extensive homology with all of the tyrosine protein kinases (HAMPE et al. 1984) and preliminary data suggest that the presumptive product of the c-*fms* gene possesses tyrosine protein kinase activity (RETTENMEIER et al. 1985).

The v-*fms* gene is expressed in the form of a *gag-fms* precursor glycoprotein termed gp180$^{gag\text{-}fms}$ (BARBACID et al. 1980b; RUSCETTI et al. 1980; VAN DE VEN et al. 1980b). Proteolytic processing yields a mature product, gp140^{fms}, which lacks most or all of the *gag*-derived sequences (ANDERSON et al. 1984). gp140^{fms} is a cell surface glycoprotein which traverses the plasma membrane once (ROUSSEL et al. 1984) (MANGER et al. 1985). The extracellular glycosylated domain is composed of approximately 550 amino acids and contains 12 potential glycosylation sites (HAMPE et al. 1984). The cytoplasmic domain is comprised of approximately 400 residues and contains the presumptive catalytic domain. A single 26 residue membrane-spanning segment separates the extracellular and intracellular domains. gp140^{fms} therefore resembles gp74^{erbB}, the EGF receptor and the insulin receptor in both structure and cellular location.

10.1 Phosphorylation

The uncertainty as to whether *fms* encodes a tyrosine protein kinase has as its origin the fact that gp140^{fms} does not always exhibit tyrosine protein kinase activity in vitro and contains little if any phosphotyrosine when isolated from transformed cells (BARBACID and LAUVER 1981; REYNOLDS et al. 1981). Additionally, evidence for cellular proteins which are newly-phosphorylated on tyrosine in McDonough FeSV-transformed cells is lacking (BARBACID and LAUVER 1981; REYNOLDS et al. 1981). These somewhat contradictory properties are not totally dissimilar to what has been observed with both *erb*B and *ros* (see above) and most probably indicate that detection of the tyrosine protein kinase activity of the *fms* product is simply not as straight-forward as is the study of *src* or *abl*.

10.2 c-fms

The c-*fms* gene is expressed in placenta, myeloid cells and the spleen (MÜLLER et al. 1982; MÜLLER et al. 1983; RETTENMEIER et al. 1985). Recent results suggest that the c-*fms* gene encodes a 170000 dalton protein which is closely related to the colony-stimulating Factor I receptor of macrophages (RETTENMEIER et al. 1985).

11 General Properties of the Viral Tyrosine Protein Kinases

A few generalizations about the viral tyrosine protein kinases can be made. Without exception, they act in the cytoplasm of infected cells. A majority of the viral kinases are present at the cytoplasmic face of the plasma membrane. It is reasonable to expect that at least some of their critical primary targets also will be found here. Not all viral tyrosine protein kinases are bona fide membrane proteins. If the *fps*-containing transforming proteins are bound to cellular membranes, they are bound only tenuously.

As a rule, the viral tyrosine protein kinases are themselves the substrates of protein kinases. Most contain both phosphoserine and phosphotyrosine, although this has not yet been demonstrated conclusively for either gp74^{erbB} and P68^{ros}. Just as it is likely that the phosphorylation of cellular proteins by the viral kinases affects their activity, it is reasonable to suspect that the activity of the kinases themselves is also regulated by phosphorylation. There are strong indications that the phosphorylation of p60$^{v\text{-}src}$, p60$^{c\text{-}src}$, and P140^{fps} on tyrosine increases either their protein kinase activity, their ability to induce transformation, or their ability to induce tumors (CROSS and HANAFUSA 1983; SNYDER and BISHOP 1984; WEINMASTER et al. 1984; COLLETT et al. 1984; BROWN and GORDON 1984). Whether this is a property of all members of this class of transforming protein is a question which it is now important to answer.

12 The Evolution and Divergence of the Genes for the Cellular Tyrosine Protein Kinases

The genomes of higher vertebrates apparently contain the genes for at least 12 protein kinases with a specificity for tyrosine. Comparison of the known sequences of the proteins encoded by these genes shows that they all possess homologous catalytic domains. There seems no doubt that these comprise a divergent gene family.

Divergence must have occurred early in evolution. *Drosophila* possess and express the genes for at least three clearly distinct tyrosine protein kinases (HOFFMAN et al. 1983; SIMON et al. 1983; LEV et al. 1984; LIVNEH et al. 1985; WADSWORTH et al. 1985). DNA sequence analysis of the *Drosophila* homologues of the c-*src*, c-*abl*, and EGF receptor genes has found that the differing structures and sequences of these three genes were established at least as early as the time of the evolution of *Drosophila* and that they have been maintained as distinct genetic entities since then.

Why do vertebrate genomes encode such a number of tyrosine protein kinases? One possibility is that the products of these different genes have different polypeptide substrate specificities. There is, however, little evidence to date that this is the case. The spectrum of cellular proteins phosphorylated on tyrosine in cells transformed by the different viral tyrosine protein kinases is remarkably similar.

Another possibility is that the *expression* of each of these several genes with apparently similar function is regulated differently. There is some evidence that this is the case. The c-*fps* gene is expressed predominantly in lymphoid cells (SHIBUYA et al. 1982), the c-*abl* gene is expressed at an unusually high level in the testes (MÜLLER et al. 1982), and the expression of both the c-*src* and c-*abl* genes is carefully regulated during *Drosophila* embryogenesis (LEV et al. 1984).

A third possibility is that divergence has given rise to tyrosine protein kinases possessing similar substrate specificities, but subject to different modes by which the activity of the protein is regulated. For example, the protein kinase activity of the cytoplasmic catalytic domain of the EGF receptor is obvious only after the binding of EGF to the extracellular amino terminal domain of the protein. In contrast, $p60^{c\text{-}src}$ exhibits apparently constitutive tyrosine protein kinase activity and possesses no identified regulatory domain.

13 The Future

The realization that the *sis* oncogene was, in fact, the gene for one of the subunits of platelet-derived growth factor provided the first conclusive link between viral oncogenes and growth factor and their receptors (DOOLITTLE et al. 1983). This observation suggested strongly that chronic activation of the receptor for a growth factor could induce transformation. This notion was strengthened enormously by the subsequent realization that the *erb*B oncogene encoded a fragment of the growth factor receptor for EGF (DOWNWARD et al. 1984a). It is unlikely that these will prove to be singular observations. New viral oncogenes which represent transduced cellular genes for other growth factors and for other growth factor receptors can be anticipated. We may well discover that one or more of the three viral oncogenes which are already known to be fragments of genes for a yet unidentified cellular tyrosine protein kinases – *fgr*, *yes*, and *ros* are in fact fragments of genes for growth factor receptors.

References

Abelson HT, Rabstein LS (1970) Lymphosarcoma: virus-induced thymic-independent disease in mice. Cancer Res 30:2213–2222

Adkins B, Hunter T, Sefton BM (1982) The transforming proteins of PRCII virus and Rous sarcoma virus form a complex with the same two cellular phosphoproteins. J Virol 43:448–455

Anderson SJ, Gonda MA, Rettenmier CW, Sherr CJ (1984) Subcellular localization of glycoproteins encoded by the viral oncogene v-*fms*. J Virol 51:730–741

Antler AM, Greenberg ME, Edelman GM, Hanafusa H (1985) Increased phosphorylation of tyrosine in vinculin does not occur upon transformation by some avian sarcoma viruses. Mol Cell Biol 5:263–267

Balduzzi PC, Notter MFD, Morgan HR, Shibuya M (1981) Some biological properties of two new avian sarcoma viruses. J Virol 40:268–275

Barbacid M, Lauver AV (1981) Gene products of McDonough feline sarcoma virus have an in vitro-associated protein kinase that phosphorylates tyrosine residues: lack of detection of this enzymatic activity in vivo. J Virol 40:812–821

Barbacid M, Beemon K, Devare SG (1980a) Origin and functional properties of the major gene product of Snyder-Theilen strain of feline sarcoma virus. Proc Natl Acad Sci USA 77:5158–5162

Barbacid ML, Lauver AV, Devare SG (1980b) Biochemical and immunological characterization of polyproteins coded for by the McDonough, Gardner-Arnstein, and Snyder-Theilen strains of feline sarcoma virus. J Virol 33:196–207

Besmer P, Hardy WD, Zuckerman EE, Bergold P, Lederman L, Snyder HW (1983) The Hardy-Zuckerman 2-FeSV, a new feline retrovirus with oncogene homology to Abelson-MuLV, Nature 303:825–828

Bishop JM (1983) Cellular oncogenes and retroviruses. Ann Rev Biochem 52:301–354

Bishop R, Martinez R, Nakamura KD, Weber MJ (1983) A tumor promoter stimulates phosphorylation on tyrosine. Biochem Biophys Res Commun 115:536–543

Bissell MJ, Hatie C, Calvin M (1979) Is the product of the src gene a promoter? Proc Natl Acad Sci USA 76:348–352

Blumberg PM, Driedger PE, Rossow PW (1976) Effect of a phorbol ester on a transformation-sensitive surface protein of chick fibroblasts. Nature 264:446–447

Bolen JB, Thiele CJ, Israel MA, Yonemoto W, Lipsich LA, Brugge JS (1984) Enhancement of cellular *src* gene product associated tyrosyl kinase activity following polyoma virus infection and transformation. Cell 38:767–777

Breitman ML, Hirano A, Wong T, Vogt PK (1981) Characteristics of avian sarcoma virus PRCIV and comparison with strain PRCII-p. Virology 114:451–462

Bretscher A (1983) Purification of an 80000 dalton protein that is a component of the isolated microvillus cytoskeleton, and its localization in nonmuscle cells. J Cell Biol 97:425–432

Brown DJ, Gordon JA (1984) The stimulation of pp60$^{v\text{-}src}$ kinase activity by vanadate in intact cells accompanies a new phosphorylation state of the enzyme. J Biol Chem 259:9580–9586

Brugge JS, Darrow D (1984) Analysis of the catalytic domain of phosphotransferase activity of two avian sarcoma virus transforming proteins. J Biol Chem 259:4550–4557

Brugge J, Erikson E, Erikson RL (1981) The specific interaction of the Rous sarcoma virus transforming protein, pp60^{src}, and two cellular proteins. Cell 25:363–372

Brugge J, Yonemoto W, Darrow D (1983) Interaction between the Rous sarcoma virus transforming protein and two cellular phosphoproteins: analysis of the turnover and distribution of this complex. Mol Cell Biol 3:9–19

Bryant D, Parsons JT (1982) Site-directed mutagenesis of the *src* gene of Rous sarcoma virus: construction of a deletion mutant temperature sensitive for transformation. J Virol 44:683–691

Bryant D, Parsons JT (1983) Site-directed point mutation in the *src* gene of Rous sarcoma virus results in an inactive *src* gene product. J Virol 45:1211–1216

Bryant DL, Parsons JT (1984) Amino acid alterations within a highly conserved region of the Rous sarcoma virus *src* gene product pp60$^{v\text{-}src}$ inactivate tyrosine protein kinase activity. Mol Cell Biol 4:862–866

Buss JE, Sefton BM (1985) The rare fatty acid, myristic acid, is the lipid attached to the transforming protein of Rous sarcoma virus and its cellular homologue. J Virol 53:7–12

Carlberg K, Chamberlin ME, Beemon K (1984) The avian sarcoma virus PRCII lacks 1020 nucleotides of *fps* transforming gene. Virology 135:157–167

Casnellie JE, Gentry LE, Rohrschneider LR, Krebs EG (1984) Identification of the tyrosine protein kinase from LSTRA cells by the use of site-specific antibodies. Proc Natl Acad Sci USA 81:6676–6680

Castagna M, Takai Y, Kaibuchi K, Sano K, Kikkawa U, Nishizuka Y (1982) Direct activation of calcium-activated, phospholipid-dependent protein kinase by tumor-promoting phorbol esters. J Biol Chem 257:7847–7851

Cochet C, Gill GN, Meisenhelder J, Cooper JA, Hunter T (1984) C-kinase phosphorylates the EGF receptor and reduces its EGF-stimulated tyrosine protein kinase activity. J Biol Chem 259:2553–2558

Collett MS, Erikson RL (1978) Protein kinase activity associated with the avian sarcoma virus *src* gene product. Proc Natl Acad Sci USA 75:2021–2024

Collett MS, Brugge JS, Erikson RL (1979a) Characterization of a normal avian cell protein related to the avian sarcoma virus transforming gene product. Cell 15:1363–1369

Collett MS, Erikson E, Erikson RL (1979b) Structural analysis of the avian sarcoma virus transforming protein: sites of phosphorylation. J Virol 29:770–781

Collett MS, Purchio AF, Erikson RL (1980) Avian sarcoma virus transforming protein, pp60^{src}, shows protein kinase activity specific for tyrosine. Nature 285:167–169

Collett MS, Belzer SK, Purchio AF (1984) Structurally and functionally modified forms of pp60$^{v\text{-}src}$ in Rous sarcoma virus-transformed cell lysates. Mol Cell Biol 4:1213–1220

Collins SJ, Groudine MT (1983) Rearrangement and amplification of c-*abl* sequences in human chronic myelogenous leukemia cell line K-562. Proc Natl Acad Sci USA 80:4813–4817

Cooper JA, Hunter T (1981a) Changes in protein phosphorylation in Rous sarcoma virus transformed chicken embryo cells. Mol Cell Biol 1:165–178

Cooper JA, Hunter T (1981b) Four different classes of retroviruses induce phosphorylation of tyrosines present in similar cellular proteins. Mol Cell Biol 1:394–407

Cooper JA, Hunter T (1981c) Similarities and differences between the effects of epidermal growth factor and Rous sarcoma virus. J Cell Biol 91:878–883

Cooper JA, Hunter T (1982) Discrete primary locations of a tyrosine protein kinase and of three proteins that contain phosphotyrosine in virally transformed chick fibroblasts. J Cell Biol 94:287–296

Cooper JA, Hunter T (1984) Regulation of cell growth and transformation by tyrosine-specific protein kinases: the search for important cellular substrate proteins. Curr Top Microbiol Immunol 107:125–162

Cooper JA, Bowen-Pope D, Raines E, Ross R, Hunter T (1982) Similar effects of platelet-derived growth factor and epidermal growth factor on the phosphorylation of tyrosine in cellular proteins. Cell 31:263–273

Cooper JA, Reiss NA, Schwartz RJ, Hunter T (1983) Three glycolytic enzymes are phosphorylated at tyrosine in cells transformed by Rous sarcoma virus. Nature 302:218–223

Cooper JA, Sefton BM, Hunter T (1984) Diverse mitogenic agents induce the phosphorylation of two related 42000 dalton proteins on tyrosine in quiescent chick cells. Mol Cell Biol 4:30–37

Cooper JA, Hunter T, Shalloway D (1985) Protein-tyrosine kinase activity of pp60$^{c\text{-}src}$ is restricted in intact cells. In: Feramisco J, Stiles C, Ozanne E (eds) Cancer Cells 3. Growth factors and transformation. Cold Spring Harbory Laboratory, Cold Spring Harbor, New York

Cotton PC, Brugge JS (1983) Neural tissues express high levels of the cellular *src* gene product pp60$^{c\text{-}src}$. Mol Cell Biol 3:1157–1162

Courtneidge SA, Bishop JM (1982) Transit of pp60$^{v\text{-}src}$ to the plasma membrane. Proc Natl Acad Sci USA 79:7117–7121

Courtneidge SA, Smith AE (1983) Polyoma virus transforming protein associates with the product of the c-*src* cellular gene. Nature 303:435–439

Courtneidge SA, Levinson AD, Bishop JM (1980) The protein encoded by the transforming gene of avian sarcoma virus (pp60^{src}) and a homologous protein in normal cells (pp60$^{proto\text{-}src}$) are associated with the membrane. Proc Natl Acad Sci USA 77:3783–3787

Courtneidge SA, Ralston R, Alitalo K, Bishop JM (1983) The subcellular location of an abundant substrate (p36) for tyrosine-specific protein kinases. Mol Cell Biol 3:340–350

Cross FR, Hanafusa H (1983) Local mutagenesis of Rous sarcoma virus: the major sites of tyrosine and serine phosphorylation of p60^{src} are dispensable for transformation. Cell 34:597–607

Cross FR, Garber EA, Pellman D, Hanafusa H (1984) A short sequence in the p60^{src} N terminus is required for p60^{src} myristylation and membrane association, and for cell transformation. Mol Cell Biol 4:1834–1842

Czernilofsky AP, Levinson AD, Varmus HE, Bishop JM, Tischler E, Goodman HM (1980) Nucleotide sequence of an avian sarcoma virus oncogene (*src*) and proposed amino acid sequence for the gene product. Nature 287:193–203

Czernilofsky AP, Levinson AD, Varmus HE, Bishop JM, Tischler E, Goodman HM (1983) Corrections to the nucleotide sequence of the *src* gene of Rous sarcoma virus. Nature 301:736–738

Decker SJ (1985) Phosphorylation of the *erb*B gene product from avian erythroblastosis virus transformed chick fibroblasts. J Biol Chem 260:2003–2006

Doolittle RK, Hunkapillar MW, Hood LE, Devare SG, Robbins KC, Aaronson SA, Antoniades HN (1983) Simian sarcoma virus *onc* gene, v-*sis*, is derived from the gene (or genes) encoding a platelet-derived growth factor. Science 221:275–277

Downward J, Yarden Y, Mayes E, Scrace G, Totty N, Stockwell P, Ullrich A, Schlessinger J, Waterfield MD (1984a) Close similarity of epidermal growth factor receptor and v-*erb*-B oncogene protein sequences. Nature 307:521–527

Downward J, Parker P, Waterfield MD (1984b) Autophosphorylation sites on the receptor for epidermal growth factor. Nature 311:483–485

Duesberg PH, Phares W, Lee W-H (1983) The low tumorigenic potential of PRCII, among viruses of the Fujinami sarcoma virus subgroup, corresponds to an internal (*fps*) deletion of the transforming gene. Virology 131:144–158

Ek B, Westermark B, Wasteson A, Heldin C-H (1982) Stimulation of tyrosine-specific phosphorylation by platelet-derived growth factor. Nature 295:419–420

Erikson E, Erikson RL (1980) Identification of a cellular protein substrate phosphorylated by the avian sarcoma virus transforming gene product. Cell 21:829–836

Fabricant RN, De Larco JE, Todaro GJ (1977) Nerve growth factor receptors on human melanoma cells in culture. Proc Natl Acad Sci USA 74:565–569

Feldman RA, Wang L-H, Hanafusa H, Balduzzi PC (1982) Avian sarcoma virus UR2 encodes a transforming protein which is associated with a unique protein kinase activity. J Virol 42:228–236

Feldman RA, Wang E, Hanafusa H (1983) Cytoplasmic localization of the transforming protein of Fujinami sarcoma virus: Salt-sensitive association with subcellular components. J Virol 45:782–791

Foster DA, Hanafusa H (1983) A *fps* gene without *gag* sequences transforms cells in culture and induces tumors in chickens. J Virol 48:744–751

Friedman B, Frackelton AR, Ross AH, Connors JM, Fujiki H, Sugimura T, Rosner MR (1984) Tumor promoters block tyrosine specific phosphorylation of the epidermal growth factor receptor. Proc Natl Acad Sci USA 81:3034–3038

Garber EA, Kreuger JG, Goldberg AR (1982) Novel localization of pp60^{src} in Rous sarcoma virus-transformed rat and goat cells and in chicken cells transformed by viruses rescued from these mammalian cells. Virology 118:419–429

Geiger B (1979) A 130K protein from chicken gizzard: its location at the termini of microfilament bundles in cultured chicken cells. Cell 18:193–205

Gentry LE, Rohrschneider LR (1984) Common features of the *yes* and *src* gene products defined by peptide-specific antibodies. J Virol 51:539–546

Ghysdael J, Neil JC, Vogt PK (1981) A third class of avian sarcoma virus, defined by related transformation-specific proteins of Yamaguchi 73 and Esh sarcoma viruses. Proc Natl Acad Sci USA 78:2611–2615

Gilmore T, Martin GS (1983) Phorbol ester and diacylglycerol induce protein phosphorylation at tyrosine. Nature 306:487–490

Gilmore T, DeClue JE, Martin GS (1985) Protein phosphorylation at tyrosine is induced by the v-*erb*B gene product in vivo and in vitro. Cell 40:609–618

Groffen J, Heisterkamp N, Shibuya M, Hanafusa H, Stephenson JR (1983) Transforming genes of avian (v-*fps*) and mammalian (v-*fes*) retroviruses correspond to a common cellular locus. Virology 125:480–486

Guyden JC, Martin GS (1982) Transformation parameters of chick embryo fibroblasts transformed by Fujinami, PRCII, PRCII-p, and Y73 avian sarcoma viruses. Virology 122:71–83

Hampe A, Laprevotte I, Galibert F, Fedele LA, Sherr CJ (1982) Nucleotide sequences of feline retroviral oncogenes (v-*fes*) provide evidence for a family of tyrosine-specific protein kinase genes. Cell 30:775–785

Hampe A, Gobet M, Scherr CJ, Galibert F (1984) Nucleotide sequence of the feline retroviral oncogene v-*fms* shows unexpected homology with oncogenes encoding tyrosine-specific protein kinases. Proc Natl Acad Sci USA 81:85–89

Hanafusa T, Wang L-H, Anderson SM, Karess RE, Hayward WS, Hanafusa H (1980) Characterization of the transforming gene of Fujinami sarcoma virus. Proc Natl Acad Sci USA 77:3009–3013

Hayman MJ, Beug H (1984) Identification of a form of the avian erythroblastosis virus *erb*-B gene product at the cell surface. Nature 309:460–462

Hayman MJ, Ramssay GM, Savin K, Kitchner G, Graf T, Beug H (1983) Identification and characterization of the avian erythroblastosis virus *erb*B gene product as a membrane glycoprotein. Cell 32:579–588

Heisterkamp N, Stephenson JR, Groffen J, Hansen PF, de Klein A, Bartram CR, Grosfeld G (1983) Localization of the c-*abl* oncogene adjacent to a translocation point in chronic myelocytotic leukemia. Nature 306:239–242

Hoffman FM, Fresco LD, Hoffmann-Falk H, Shilo B-Z (1983) Nucleotide sequence of the Drosophila *src* and *abl* homologs: conservation and variability in the *src* family oncogenes. Cell 35:393–340

Hunter T, Cooper JA (1983) The role of tyrosine phosphorylation in malignant transformation and in cellular growth control. Prog Nucl Acid Res Mol Biol 29:221–232

Hunter T, Sefton BM (1980) The transforming gene product of Rous sarcoma virus phosphorylates tyrosine. Proc Natl Acad Sci USA 77:1311–1315

Hunter T, Ling N, Cooper JA (1984) C-kinase phosphorylates the EGF receptor at a threonine nine residues from the cytoplasmic face of the plasma membrane. Nature 311:480–483

Iba H, Takeya T, Cross FR, Hanafusa T, Hanafusa H (1984) Rous sarcoma virus variants that carry the cellular *src* gene instead of the viral *src* gene cannot transform chicken embryo fibroblasts. Proc Natl Acad Sci USA 81:4424–4428

Iwashita S, Fox CF (1984) Epidermal growth factor and potent phorbol tumor promoters induce epidermal growth factor receptor phosphorylation in a similar but distinctively different manner in human epidermoid carcinoma A431 cells. J Biol Chem 259:2559–2567

Jacobs S, Kull FC, Earp HS, Svoboda ME, Van Wyck JJ, Cuatrecasas P (1983) Somatomedin-C stimulates the phosphorylation of the -subunit of its own receptor. J Biol Chem 258:9581–9584

Kamps MP, Taylor SS, Sefton BM (1984) Oncogenic tyrosine protein kinases and cAMP-dependent protein kinase have homologous ATP binding sites. Nature 310:589–592

Kamps MP, Buss JE, Sefton BM (1985) Mutation of N-terminal glycine of $p60^{src}$ prevents both fatty acylation and morphological transformation. Proc Natl Acad Sci USA 82:4625–4628

Karess RE, Hanafusa H (1981) Viral and cellular *src* genes contribute to the structure of recovered avian sarcoma virus transforming protein. Cell 24:155–164

Kitamura N, Yoshida M (1983) Small deletion in *src* of RSV modifying transformation phenotypes: Identification of 207 nucleotide deletion and its smaller product with protein kinase activity. J Virol 46:985–992

Kitamura N, Kitamura A, Toyoshima K, Hirayama Y, Yoshida M (1982) Avian sarcoma virus Y73 genome sequence and structural similarity of its transforming gene product to that of Rous sarcoma virus. Nature 297:205–208

Konopka JB, Watanabe SM, Witte ON (1984) An alteration of the human c-*abl* protein in K562 leukemia cells unmasks associated tyrosine kinase activity. Cell 37:1035–1042

Kreuger JG, Wang E, Garber EA, Goldberg AR (1980a) Differences in intracellular location of $pp60^{src}$ in rat and chicken cells transformed by Rous sarcoma virus. Proc Natl Acad Sci USA 77:4142–4146

Kreuger JG, Wang E, Goldberg AR (1980b) Evidence that the *src* gene product of Rous sarcoma virus is membrane-associated. Virology 101:25–40

Kreuger JG, Garber EA, Goldberg AR (1983) Subcellular localization of $pp60^{src}$ in RSV-transformed cells. Curr Top Microbiol Immunol 107:52–124

Kris RM, Lax I, Gullick W, Waterfield MD, Ullrich A, Fridkin M, Schlessinger J (1985) Antibodies against a synthetic peptide as a probe for the kinase activity of the avian EGF receptor and the v-*erb*B protein. Cell 40:619–625

Krzyzek RA, Mitchell RL, Lau AF, Faras AJ (1980) Association of $pp60^{src}$ and *src* protein kinase activity with the plasma membrane of non-permissive and permissive avian sarcoma virus-infected cells. J Virol 36:805–815

Lee W-H, Bister K, Pawson A, Robins T, Moscovici C, Duesberg PH (1980) Fujinami sarcoma virus: an avian RNA tumor virus with a unique transforming gene. Proc Natl Acad Sci USA 77:2018–2022

Lev Z, Leibovitz N, Segev O, Shilo B-Z (1984) Expression of the *src* and *abl* cellular oncogenes during development of *Drosophila melanogaster*. Mol Cell Biol 4:982–984

Levinson AD, Oppermann H, Levintow L, Varmus HE, Bishop JM (1978) Evidence that the transforming gene of avian sarcoma virus encodes a protein kinase associated with a phosphoprotein. Cell 15:561–572

Levinson AD, Courtneidge SA, Bishop JM (1981) Structural and functional domains of the Rous sarcoma virus transforming protein ($pp60^{src}$). Proc Natl Acad Sci USA 78:1624–1628

Lipsich LA, Cutt JR, Brugge JS (1982) Association of the transforming proteins of Rous, Fujinami, and Y73 avian sarcoma viruses with the same two cellular proteins. Mol Cell Biol 2:875–880

Livneh E, Glazer L, Segal D, Schlessinger J, Shilo B-Z (1985) The *Drosophila* EGF receptor gene homolog: conservation of both hormone binding and kinase domains. Cell 40:599–607

Macara IG, Marinetti GV, Balduzzi PC (1984) Transforming protein of avian sarcoma virus UR2 is associated with phosphatidylinositol kinase activity: possible role in tumorigenesis. Proc Natl Acad Sci USA 81:2728–2732

Manger R, Najita L, Nichols EJ, Hakomori S-I, Rohrschneider L (1984) Cell surface expression of the McDonough strain of feline sarcoma virus *fms* gene product (gp140^{fms}). Cell 39:327–337

Mathey-Prevot B, Hanafusa H, Kawai S (1982) A cellular protein is immunologically cross-reactive with and functionally homologous to the Fujinami sarcoma virus transforming protein. Cell 28:897–906

Moss P, Radke K, Carter C, Young J, Gilmore T, Martin GS (1984) Cellular localization of the transforming protein of wild-type and temperature-sensitive Fujinami sarcoma virus. J Virol 52:557–565

Müller R, Slamon DJ, Tremblay JM, Cline MJ, Verma IM (1982) Differential expression of cellular oncogenes during pre- and postnatal development of the mouse. Nature 299:640–644

Müller R, Slamon DJ, Adamson ED, Tremblay JM, Meuller D, Cline MJ, Verma IM (1983) Transcription of c-*onc* genes c-*ras*^{Ki} and c-*fms* during mouse development. Mol Cell Biol 3:1062–1069

Naharro G, Dunn CY, Robbins KC (1983) Analysis of the primary translation product and integrated DNA of a new feline sarcoma virus, GR-FeSV. Virology 125:502–507

Naharro G, Robbins KC, Reddy EP (1984) Gene product of v-*fgr onc*: hybrid protein containing a portion of actin and a tyrosine-specific protein kinase. Science 223:63–66

Nakamura KD, Martinez R, Weber MJ (1983) Tyrosine phosphorylation of specific proteins following mitogen stimulation of chicken embryo fibroblasts. Mol Cell Biol 3:380–390

Neckameyer WS, Wang L-H (1985) Nucleotide sequence of avian sarcoma virus UR2 and comparison of its transforming gene with other members of the tyrosine protein kinase oncogene family. J Virol 53:879–884

Neel BG, Wang L-H, Mathey-Prevot B, Hanafusa T, Hanafusa H, Hayward WS (1982) Isolation of 16L virus: a rapidly transforming sarcoma virus from an avian leukosis virus-induced sarcoma. Proc Natl Acad Sci USA 79:5088–5092

Neil JC, Ghysdael J, Vogt PK (1981a) Tyrosine-specific protein kinase activity associated with p105 of avian sarcoma virus PRCII. Virology 109:223–228

Neil JC, Ghysdael J, Vogt PK, Smart JE (1981b) Homologous tyrosine phosphorylation sites in transformation-specific products of distinct avian sarcoma viruses. Nature 291:675–677

Nigg EA, Cooper JA, Hunter H (1983) Immunofluorescent localization of a 39000 dalton substrate of tyrosine protein kinases to the cytoplasmic surface of the plasma membrane. J Cell Biol 97:1601–1609

Nishimura J, Huang JS, Deuel TF (1982) Platelet-derived growth factor stimulates tyrosine-specific protein kinase activity in Swiss mouse 3T3 cell membranes. Proc Natl Acad Sci USA 79:4303–4307

Nishizuka Y (1983) Phospholipid degradation and signal translation for protein phosphorylation. Trends Biochem Sci 8:13–16

Notter MFD, Balduzzi PC (1984) Cytoskeletal changes induced by two avian sarcoma viruses: UR2 and Rous sarcoma virus. Virology 136:56–68

Oppermann H, Levinson AD, Levintow L, Varmus HE, Bishop JM (1979) Uninfected vertebrate cells contain a protein that is closely related to the product of the avian sarcoma virus transforming gene (*src*). Proc Natl Acad Sci USA 76:1804–1808

Oppermann H, Levinson AD, Levintow L, Varmus HE, Bishop JM, Kawai S (1981) Two cellular proteins that immunoprecipitate with the transforming protein of Rous sarcoma virus. Virology 113:736–751

Parker RC, Varmus HE, Bishop JM (1984) Expression of v-*src* and chicken c-*src* in rat cells demonstrates qualitative differences between pp60$^{v\text{-}src}$ and pp60$^{c\text{-}src}$. Cell 37:131–139

Parsons JT, Bryant D, Wilkerson V, Gilmartin G, Parsons SJ (1984) Site-directed mutagenesis of Rous sarcoma virus pp60$^{v\text{-}src}$: identification of functional domains required for transformation. Cancer cells, vol 2: oncogenes and viral genes. Cold Spring Harbor Laboratory, New York, pp 37–42

Patschinsky T, Sefton BM (1981) Evidence that there exist four classes of RNA tumor viruses which encode proteins with associated tyrosine protein kinase activities. J Virol 39:104–114

Patschinsky T, Hunter T, Esch FS, Cooper JA, Sefton BM (1982) Analysis of the sequence of amino acids surrounding sites of tyrosine phosphorylation. Proc Natl Acad Sci USA 79:973–977

Pawson T, Guyden J, Kung T-H, Radke K, Gilmore T, Martin GS (1980) A strain of Fujinami sarcoma virus which is temperature sensitive in protein phosphorylation and cellular transformation. Cell 22:767–776

Ponticelli AS, Whitlock CA, Rosenberg N, Witte ON (1982) In vivo tyrosine phosphorylations of the Abelson virus transforming protein are absent in its normal cellular homolog. Cell 29:953–960

Privalsky MS, Sealy L, Bishop JM, McGrath JP, Levinson A (1983) The product of the avian erythroblastosis virus *erb*B genlocus is a glycoprotein. Cell 32:1257–1267

Prywes R, Foulkes JG, Rosenberg N, Baltimore D (1983) Sequences of the A-MuLV protein needed for fibroblast and lymphoid cell transformation. Cell 34:569–579

Prywes R, Hoag J, Rosenberg N, Baltimore D (1985) Protein stabilization explains the *gag* requirement for transformation of lymphoid cells by Abelson murine leukemia virus. J Virol 54:123–132

Purchio AF (1982) Evidence that $pp60^{src}$, the product of the Rous sarcoma virus *src* gene, undergoes autophosphorylation. J Virol 41:1–7

Purchio AF, Erikson E, Brugge JS, Erikson RL (1978) Identification of a polypeptide encoded by the avian sarcoma virus *src* gene. Proc Natl Acad Sci USA 75:1567–1571

Radke K, Martin GS (1979) Transformation by Rous sarcoma virus: Effects of *src* gene expression on the synthesis and phosphorylation of cellular polypeptides. Proc Natl Acad Sci USA 76:5212–5216

Radke K, Gilmore T, Martin GS (1980) Transformation by Rous sarcoma virus: a cellular substrate for transformation-specific protein phosphorylation contains phosphotyrosine. Cell 21:821–828

Radke K, Carter VC, Moss P, Dehazya P, Schliwa M, Martin GS (1983) Membrane association of a 36000 dalton substrate for tyrosine phosphorylation in chicken embryo fibroblasts transformed by avian sarcoma viruses. J Cell Biol 97:1601–1611

Reddy EP, Smith MJ, Srinivasan A (1983) Nucleotide sequence of Abelson murine leukemia virus genome: Structural similarity of its transforming gene product to other *onc* gene products with tyrosine-specific kinase activity. Proc Natl Acad Sci USA 80:3623–3627

Resh MD, Erikson RL (1985) Highly specific antibody to Rous sarcoma virus transforming protein recognizes a novel population of $pp60^{src}$ molecules. J Cell Biol (to be published)

Rettenmier CW, Chen JH, Roussel MF, Sherr CJ (1985) The product of the c-*fms* oncogene: a glycoprotein with associated tyrosine kinase activity. Science 228:320–322

Reynolds FH, Van de Ven WJM, Blomberg J, Stephenson JR (1981) Differences in mechanisms of transformation by independent feline sarcoma virus isolates. J Virol 38:1084–1089

Reynolds FH, Oroszlan S, Stephenson JR (1982) Abelson murine leukemia virus P120: identification and characterization of tyrosine phosphorylation sites. J Virol 44:1097–1101

Rifkin DB, Crowe RM, Pollack R (1979) Tumor promoters induce changes in the chick embryo fibroblast cytoskeleton. Cell 18:361–368

Rohrschneider LR (1980) Adhesion plaques of Rous sarcoma virus transformed cells contain the *src* gene product. Proc Natl Acad Sci USA 77:3514–3518

Rohrschneider LR, Najita LM (1984) Detection of the v-*abl* gene product at cell-substratum contact sites in Abelson murine leukemia virus-transformed fibroblasts. J Virol 51:547–552

Rosenberg N, Witte ON (1980) Abelson murine leukemia virus mutations with alterations in the virus-specific P120 molecule. J Virol 33:340–348

Rosenberg N, Baltimore D, Scher CD (1975) In vitro transformation of lymphoid cells by Abelson murine leukemia virus. Proc Natl Acad Sci USA 72:1932–1936

Rosenberg N, Clark DR, Witte ON (1980) Abelson murine leukemia virus mutants deficient in kinase activity and lymphoid cell transformation. J Virol 36:766–774

Roth CW, Richert ND, Pastan I, Gottesman MM (1983) Cyclic AMP treatment of Rous sarcoma virus-transformed chinese hamster ovary cells increases phosphorylation of $pp60^{src}$ and increases $pp60^{src}$ kinase activity. J Biol Chem 258:10768–10773

Roussel MR, Rettenmeier CW, Look AT, Sherr CJ (1984) Cell surface expression of v-*fms*-coded glycoproteins is required for transformation. Mol Cell Biol 4:1999–2009

Rozengurt E, Rodriguez-Pena A, Coombs M, Sinnett-Smith J (1984) Diacylglycerol stimulates DNA synthesis and cell division in mouse 3T3 cells: role of Ca^{2+}-sensitive phospholipid-dependent protein kinase. Proc Natl Acad Sci USA 81:5748–5752

Rubin JB, Shia MA, Pilch PF (1983) Stimulation of tyrosine-specific phosphorylation in vitro by insulin-like growth factor I. Nature 305:438–440

Ruscetti SK, Turek LP, Sherr CJ (1980) Three independent isolates of feline sarcoma virus code for distinct *gag*-X polyproteins. J Virol 35:259–264

Scher CD, Siegler R (1975) Direct transformation of 3T3 cells by Abelson murine leukemia virus. Nature 253:729–731

Schultz AM, Oroszlan S (1983) In vivo modification of retroviral gag gene-encoded polyproteins by myristic acid. J Virol 46:355–361

Schultz AM, Oroszlan S (1984) Myristylation of *gag-onc* fusion proteins in mammalian transforming retroviruses. Virology 133:431–437

Schulz AM, Henderson LE, Oroszlan S, Garber EA, Hanafusa H (1985) Amino terminal myristylation of the protein kinase p60*src*, a retroviral transforming protein. Science 227:427–429

Schwartz D, Tizard R, Gilbert W (1983) Nucleotide sequence of Rous sarcoma virus. Cell 32:853–869

Sefton BM, Hunter T (1984) Tyrosine protein kinases. Adv Cyclic Nucleotide Protein Phosphorylation Res 18:195–226

Sefton BM, Hunter T, Beemon K, Eckhart W (1980) Phosphorylation of tyrosine is essential for cellular transformation by Rous sarcoma virus. Cell 20:807–816

Sefton BM, Hunter T, Raschke WC (1981a) Evidence that the Abelson virus protein functions in vivo as a protein kinase which phosphorylates tyrosine. Proc Natl Acad Sci USA 78:1552–1556

Sefton BM, Hunter T, Ball EH, Singer SJ (1981b) Vinculin: a cytoskeletal substrate of the transforming protein of Rous sarcoma virus. Cell 24:165–174

Sefton BM, Patschinsky T, Berdot C, Hunter T, Elliott T (1982) Phosphorylation and metabolism of the transforming protein of Rous sarcoma virus. J Virol 41:813–820

Sefton BM, Hunter T, Cooper JA (1983) Some lymphoid cell lines transformed by Abelson murine leukemia virus lack a major 36000 dalton tyrosine protein kinase substrate. Mol Cell Biol 3:56–63

Shalloway D, Coussens PM, Yaciuk P (1984) Overexpression of the c-*src* protein does not induce transformation of NIH 3T3 cells. Proc Natl Acad Sci USA 81:7071–7075

Shibuya M, Hanafusa H (1982) Nucleotide sequence of Fujinami sarcoma virus: Evolutionary relationship of its transforming gene with transforming genes of other sarcoma viruses. Cell 30:787–795

Shibuya M, Hanafusa T, Hanafusa H, Stephenson JR (1980) Homology exists among the transforming sequences of avian and feline sarcoma viruses. Proc Natl Acad Sci USA 77:6536–6540

Shibuya M, Hanafusa H, Balduzzi PC (1982) Cellular sequences related to three new onc genes of avian sarcoma virus (*fps*, *yes*, and *ros*) and their expression in normal and transformed cells. J Virol 42:143–152

Shields A, Otto G, Goff S, Baltimore D (1979) Structure of the Abelson murine leukemia virus genome. Cell 18:955–962

Simon MA, Kornberg T, Bishop JM (1983) Three loci related to the *src* oncogene and tyrosine-specific protein kinase activity in *Drosophila*. Nature 302:837–839

Smart JE, Oppermann H, Czernilofsky AP, Purchio AF, Erikson RL, Bishop JM (1981) Characterization of sites for tyrosine phosphorylation in the transforming protein of Rous sarcoma virus (pp60^{src}) and its normal cellular homologue (pp60$^{c\text{-}src}$). Proc Natl Acad Sci USA 78:6013–6017

Snyder MA, Bishop JM (1984) A mutation at the major phosphotyrosine in pp60$^{v\text{-}src}$ alters oncogenic potential. Virology 136:375–386

Snyder MA, Bishop JM, Colby WW, Levinson AD (1983) Phosphorylation of tyrosine-416 is not required for the transforming properties and kinase activity of pp60$^{v\text{-}src}$. Cell 32:891–901

Sorge LK, Levey BT, Maness PF (1984) pp60$^{c\text{-}src}$ is developmentally regulated in the neural retina. Cell 36:249–257

Stehelin D, Varmus HE, Bishop JM, Vogt PK (1976) DNA related to the transforming gene(s) of avian sarcoma viruses is present in normal avian DNA. Nature 260:170–173

Sugimoto Y, Whitman M, Cantley LC, Erikson RL (1984) Evidence that the Rous sarcoma virus transforming gene product phosphorylates phosphatidylinositol and diacylglycerol. Proc Natl Acad Sci USA 81:2117–2121

Swanstrom R, Parker RC, Varmus HE, Bishop JM (1983) Transduction of a cellular oncogene: the genesis of Rous sarcoma virus. Proc Natl Acad Sci USA 80:2519–2523

Takeya T, Hanafusa H (1983) Structure and sequence of the cellular gene homologous to the RSV *src* gene and the mechanism for generating the transforming virus. Cell 32:881–890

Takeya T, Feldman RA, Hanafusa H (1982) DNA sequence of the viral and cellular *src* gene of chicken. I. The complete nucleotide sequence of an *Eco*RI fragment of recovered avian sarcoma virus which codes for gp37 and pp60$^{v\text{-}src}$. J Virol 44:1–11

Ullrich A, Coussens L, Hayflick JS, Dull TJ, Gray A, Tam AW, Lee J, Yarden Y, Liberman TA, Schlessinger J, Downward J, Mayes ELV, Waterfield MD, Whittle M, Seeburg PH (1984) Human epidermal growth factor receptor cDNA sequence and aberrant expression of the amplified gene in A431 epidermoid carcinoma cells. Nature 309:418–425

Ullrich A, Bell JR, Chen EY, Herrara R, Petruzzelli LM, Dull TJ, Gray A, Coussens L, Liao Y-C, Tsubokawa M, Mason A, Seeburg PH, Grunfeld C, Rosen OM, Ramachandran J (1985) Human insulin receptor and its relationship to the tyrosine kinase family of oncogenes. Nature 313:756–761

Ushiro H, Cohen S (1980) Identification of phosphotyrosine as a product of epidermal growth factor-activated protein kinase in A-431 cell membranes. J Biol Chem 255:8363–8365

Van De Ven WJM, Reynolds FH, Nalewaik RP, Stephenson JR (1980)a) Characterization of a 170000 dalton polyprotein encoded by the McDonough strain of feline sarcoma virus. J Virol 35:165–175

Van de Ven WJM, Reynolds FH, Stephenson JR (1980b) The nonstructural components of polyproteins encoded by replication-deficient mammalian transforming retroviruses are phosphorylated and have associated protein kinase activity. Virology 101:185–197

Voronova AF, Buss JE, Patschinsky T, Hunter T, Sefton BM (1985) Characterization of the protein apparently responsible for the elevated tyrosine protein kinase activity in LSTRA cells. Mol Cell Biol 4:2705–2713

Wadsworth SC, Vincent WS, Bilodeau-Wentworth D (1985) A *Drosophila* genomic sequence with homology to human epidermal growth factor receptor. Nature 314:178–180

Wang JYJ, Baltimore D (1983) Cellular RNA homologous to the Abelson murine leukemia virus transforming gene: expression and relationship to the viral sequence. Mol Cell Biol 3:773–779

Wang JYJ, Baltimore D (1985) Localization of tyrosine kinase-coding region in v-*abl* oncogene by the expression of v-*abl*-encoded proteins in bacteria. J Biol Chem 260:64–71

Wang JYJ, Ledley F, Goff S, Lee R, Groner Y, Baltimore D (1984) The mouse c-*abl* locus: molecular cloning and characterization. Cell 36:349–356

Wang L-H, Hanafusa H, Notter MFD, Balduzzi PC (1982a) Genetic structure, transforming sequence, and gene product of avian sarcoma virus UR1. J Virol 40:258–267

Wang L-H, Hanafusa H, Notter MFD, Balduzzi PC (1982b) Genetic structure and transforming sequence of avian sarcoma virus UR2. J Virol 41:833–841

Weinmaster G, Hinze E, Pawson T (1983) Mapping of multiple phosphorylation sites within the structural and catalytic domains of the Fujinami sarcoma virus transforming protein. J Virol 46:29–41

Weinmaster G, Zoller MJ, Smith M, Hinze E, Pawson T (1984) Mutagenesis of Fujinami sarcoma virus: evidence that tyrosine phosphorylation of $P130^{gag\text{-}fps}$ modulates its biological activity. Cell 37:559–568

Willingham MC, Jay G, Pastan I (1979) Localization of the ASV *src* gene product to the plasma membrane of transformed cells by immunoelectron microscopy. Cell 18:125–134

Witte ON (1983) Molecular and cellular biology of Abelson virus transformation. Curr Top Microbiol Immunol 103:127–146

Witte ON, Rosenberg N, Baltimore D (1979a) Identification of a normal cellular protein cross-reactive to the major Abelson murine leukemia virus gene product. Nature 281:396–398

Witte ON, Dasgupta A, Baltimore D (1980) Abelson murine leukemia virus protein is phosphorylated in vitro to form phosphotyrosine. Nature 283:826–831

Wong T-W, Goldberg AR (1984) Purification and characterization of the major species of tyrosine protein kinase in rat liver. J Biol Chem 259:8505–8512

Woolford J, Beemon K (1984) Transforming proteins of Fujinami and PRCII avian sarcoma viruses have different subcellular locations. Virology 135:168–180

Yamamoto T, Hihara H, Nishida T, Kawai S, Toyoshima K (1983a) A new avian erythroblastosis virus, AEV-H, carries *erb*B gene responsible for the induction of both erythroblastosis and sarcomas. Cell 34:225–234

Yamamoto T, Nishida T, Miyajima N, Kawai S, Ooi T, Toyoshima K (1983b) The *erb*B gene of avian erythroblastosis virus is a number of the *src* gene family. Cell 35:71–78

Yoshida M, Kawai S, Toyoshima K (1980) Uninfected avian cells contain structurally unrelated progenitors of viral sarcoma genes. Nature 287:653–654

Young JC, Martin GS (1984) Cellular localization of the c-*fps* gene product. J Virol 52:913–918

Homology Among Oncogenes

C. VAN BEVEREN and I.M. VERMA

1 Introduction 73
2 Genome Stategies 76
3 Oncogene Homologies 77
3.1 Tyrosine Protein Kinases 78
3.2 Oncogenes Without Manifest Tyr-PK Activity 78
3.3 *ras* Oncogenes 82
3.4 Oncogenes Whose Products Are Located in the Nucleus 83
3.5 Oncogenes Homologous to Growth Factors and Receptors 85
4 Comparison of c-*onc* and v-*onc* Products 88
5 Structural Homologies 89
6 Prospects of New Oncogenes 89
References 91

1 Introduction

At the heart of current research in cancer is the finding that genes whose products are directly implicated in oncogenesis have counterparts in normal cells (BISHOP 1983; HUNTER 1984). Such cancer genes, termed oncogenes, were first garnered via the agency of retroviruses, since they constitute an integral part of the genomes of several acutely transforming retroviruses (BISHOP and VARMUS 1982). The first viral oncogene identified was *src*, the transforming gene of Rous sarcoma virus (RSV) (STEHELIN et al. 1976). The genome of RSV contained *src* sequences in addition to the genes required for replication and virion formation. However, all other acutely transforming retroviruses identified to date are unable to support their replication, because transforming sequences are acquired at the expense of viral genes whose products are essential for propagation. Consequently all acutely transforming retroviruses are defective for replication (BISHOP and VARMUS 1982; VERMA 1983).

Oncogenic sequences present in the viral genomes are referred to as viral oncogenes (v-*onc*) and their cellular progenitors as proto-oncogenes or cellular oncogenes (c-*onc*). To date, over 20 retroviral oncogenes, each with a distinct counterpart in normal cells, have been identified and extensively analyzed (Table 1). Additionally, oncogenes have also been identified by transfection of fibroblastic cell lines with tumor DNAs (COOPER 1982; WEINBERG 1983).

Molecular Biology and Virology Laboratory, Salk Institute for Biological Studies, P.O. Box 85800, San Diego, CA 92138-9216, USA

Table 1. Catalog of oncogenes

Class	Gene	Prototype virus	Disease	Viral product	Proto-oncogene product
I	*src*	RSV	Sarcoma	$p60^{v\text{-}src}$	$p60^{c\text{-}src}$
Tyrosine protein kinases	*yes*	Y73	Sarcoma	$P90^{gag\text{-}yes}$	?
	fgr	GR-FeSV	Sarcoma	$P70^{gag\text{-}actin\text{-}fgr}$	?
	fps	FSV	Sarcoma	$P140^{gag\text{-}fps}$	p98
	fes	ST-FeSV	Sarcoma	$P90^{gag\text{-}fes}$	p92
	ros	UR2	Sarcoma	$P68^{gag\text{-}ros}$	?
	abl	Ab-MLV	Leukemia	$P160^{gag\text{-}abl}$	p150
	*erb*B	AEV	Sarcoma, erythroblastosis	$gp72^{erb\text{-}B}$	gp175
	fms	SM-FeSV	Sarcoma	$gP180^{gag\text{-}fms}$	?
IA	*mos*	Mo-MSV	Sarcoma	$P37^{env\text{-}mos}$	?
Related to	*mht/mil*	MH2	Sarcoma	$P100^{gag\text{-}mht}$	?
Tyr-PK	*raf*	3611-MSV	Sarcoma	$P75^{gag\text{-}raf}$	?
II	Ha-*ras*	Ha-MSV	Erythroleukemia	$p21^{v\text{-}Ha\text{-}ras}$	$p21^{c\text{-}Ha\text{-}ras}$
GTP-binding proteins	Ki-*ras*	Ki-MSV	Sarcoma	$p21^{v\text{-}Ki\text{-}ras}$	$p21^{c\text{-}Ki\text{-}ras}$
III	*myc*	MC29	Carcinoma, myelocytomatosis	$P110^{gag\text{-}myc}$	p58
Nuclear proteins	*myb*	AMV	Myeloblastosis	$P45^{gag\text{-}myb}$	p75
	fos	FBJ-MSV	Osteosarcoma	$p55^{v\text{-}fos}$	$p55^{c\text{-}fos}$
	ski	SKV	Carcinoma	$P110^{gag\text{-}ski}$	?
IV	*sis*	SSV	Glioma, fibrosarcoma	$P28^{env\text{-}sis}$	p30 PDGF B-chain
Growth factors and receptors	*erb*-B	AEV	Sarcoma, erythroblastosis	$gp72^{erb\text{-}B}$	gp175
V	*erb*-A	AEV-ES4	Sarcoma, erythroblastosis	$P75^{gag\text{-}erb\text{-}A}$	?
Other viral	*ets*	E26	Erythroblastosis	$P135^{gag\text{-}myb\text{-}ets}$	?
oncogenes	*rel*	REV-T	Lymphatic leukemia	$P56^{env\text{-}rel}$	?
VI	N-*ras*	–	Various	–	$p21^{N\text{-}ras}$
Nonviral	*N-myc*		Neuroblastoma	–	?
oncogenes	B*lym*		Bursal lymphoma	–	?
	neu		Neuroblastoma	–	gp180

This review is concerned with a comparison of homologies among various oncogene products. For an analysis of viral and cellular oncogene expression, and of the products of the protein kinase oncogenes, the interested reader may refer to a number of recent reviews (MÜLLER and VERMA 1984; HUNTER and COOPER 1985; SEFTON 1985).

Fig. 1A, B. Genomic strategies. The structures of various transforming retroviruses are shown with respect to the map of a replication-competent avian (**A**) or mammalian (**B**) viral genome, from capped nucleotide to polyadenylate tail. *Closed boxes* indicate regions which contribute to the LTR; *solid lines*, untranslated regions; *open boxes*, viral genes *gag*, *pol*, and *env*; *loosely hatched boxes*, v-*onc* genes with known transforming activity; *dotted closed boxes*, other cellular genes which may contribute to transformation; *tightly hatched boxes*, rat VL30 sequences; *dotted lines*, sequences deleted from the parental virus. It may be noted that the structure of the avian virus REV-T is more similar to that of MLV than to that of ALV

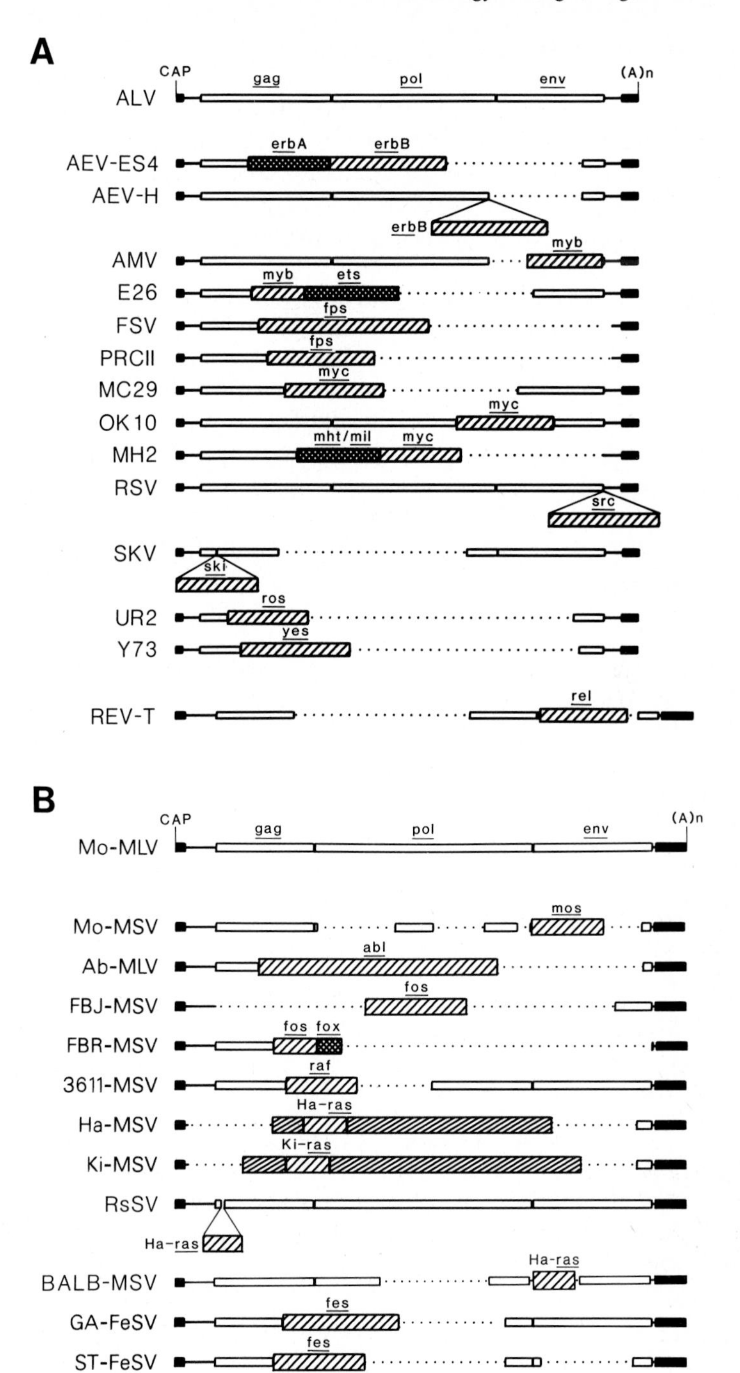
A
CAP
gag
pol
env
(A)n
ALV
AEV-ES4
erbA
erbB
AEV-H
erbB
AMV
myb
E26
myb
ets
FSV
fps
PRCII
fps
MC29
myc
OK10
myc
MH2
mht/mil
myc
RSV
src
SKV
ski
UR2
ros
Y73
yes
REV-T
rel
B
CAP
gag
pol
env
(A)n
Mo-MLV
Mo-MSV
mos
Ab-MLV
abl
FBJ-MSV
fos
FBR-MSV
fos
fox
3611-MSV
raf
Ha-MSV
Ha-ras
Ki-MSV
Ki-ras
RsSV
Ha-ras
BALB-MSV
Ha-ras
GA-FeSV
fes
ST-FeSV
fes
SM-FeSV
fms
GR-FeSV
actin
fgr
SSV
sis

2 Genome Strategies

Acutely transforming retroviruses acquire cellular sequences by recombination between replication-competent retroviruses and c-*onc* genes. While the precise mechanism of acquisition of cellular sequences remains largely obscure, there do not appear to be any specific constraints regarding the site of recombination in the coding regions of the viral genome. The retroviral genome contains various cis-acting elements, including the LTR sequences, primer-binding site, packaging signals, and the sequences required for initiation of (+) strand DNA synthesis (VARMUS and SWANSTROM 1982; VERMA 1983). No acutely transforming retroviruses have yet been discerned where recombination eliminates these cis-acting elements. The virion-encoded proteins (*gag*, *pol*, and *env* products) required for replication can be supplied in "trans" (MANN et al. 1983; MILLER et al. 1985). It is thus not surprising that recombination with cellular sequences occurs in the viral coding regions. Figure 1 outlines the structures of various acutely oncogenic retroviruses.

Retroviruses display a remarkable plasticity of the genome and consequently have evolved a large array of genomic strategies. Table 2 shows that v-*onc* gene products can be initiated in either the viral *gag* or *env* genes, or, more rarely, within the acquired cellular sequences. The product is most frequently terminated within the acquired sequences, but cases in which the product continues into viral genes have also been identified (Table 2).

Table 2. Initiation, and termination of v-*onc* Translation

Virus	Translation product[a]	Versus c-*onc* product C-terminus[b]	Sequence references
Ab-MLV	*gag-abl*	nd	REDDY et al. 1983b; R. LEE, M. PASKIND, J.Y.J. WANG, D. BALTIMORE 1984, personal communication
AEV-ES4	*gag-erb*-A-*erb*-B intron ?(*gag*)-*erb*-B-?	*erb*-A: truncated *erb*-B: nd	DEBUIRE et al. 1984; HENRY et al. 1985
AEV-H	?(*gag*)-*erb*-B-[*env*]	Truncated	YAMAMOTO et al. 1983
AMV	(*gag*)-*myb-env*	Truncated	KLEMPNAUER et al. 1982; RUSHLOW et al. 1982
BALB-MSV	Ha-*ras*	Same	REDDY et al. 1985
E26	*gag-myb-ets*	*myb*: truncated *ets*: nd	NUNN et al. 1983
FBJ-MSV	*fos*	Downstream sequences	VAN BEVEREN et al. 1983
FBR-MSV	*gag-fos-fox*	*fos*: truncated *fox*: nd	VAN BEVEREN et al. 1984
FSV	*gag-fps*	Same	SHIBUYA and HANAFUSA 1982; HUANG et al. 1985
GA-FeSV	*gag-fes*	Same? (same as ST-FeSV)	HAMPE et al. 1982
GR-FeSV	*gag*-actin-*fgr*-[*env*]	Actin: truncated *fgr*: truncated	NAHARRO et al. 1984

Table 2 (continued)

Virus	Translation product[a]	Versus c-*onc* product C-terminus[b]	Sequence references
Ha-MSV	Ha-*ras*	Same	DHAR et al. 1982; YASUDA et al. 1984
Ki-MSV	Ki-*ras*	Same	TSUCHIDA et al. 1982
MC29	?(*gag*)-*myc*	Same	ALITALO et al. 1983; REDDY et al. 1983a; WATSON et al. 1983
MH2	*gag-mht/mil* ?(*gag*)-*myc*	*mht/mil*: same? (same as 3611-MSV) *myc*: same	GALIBERT et al. 1984; KAN et al. 1984b; SUTRAVE et al. 1984
Mo-MSV	*env-mos*	Same?	REDDY et al. 1981; VAN BEVEREN et al. 1981a; DONOGHUE 1982
PRCII	*gag-fps*	Same	HUANG et al. 1984, 1985
REV-T	*env-rel*-[*env*]	Truncated	STEPHENS et al. 1983; WILHELMSEN et al. 1984
RsSV	*gag*-Ha-*ras*	Same	RASHEED et al. 1983
RSV	*src*	Downstream sequences	TAKEYA and HANAFUSA 1982; CZERNILOVSKY et al. 1983; SCHWARTZ et al. 1983
SKV	*gag-ski-gag*-[*pol*]	Truncated	E. STAVNEZER 1984, personal communication
SM-FeSV	*gag-fms*	Downstream sequences?	HAMPE et al. 1984; C. VAN BEVEREN, unpublished results
SSV	*env-sis*	Same	DEVARE et al. 1983; HANNINK and DONOGHUE 1984
ST-FeSV	*gag-fes*	Same? (same as GA-FeSV)	HAMPE et al. 1982
3611-MSV	*gag-raf*	Same? (same as MH2)	MARK and RAPP 1984
UR2	*gag-ros*	nd	NECKAMEYER and WANG 1985
Y73	*gag-yes*-[*env*]	Truncated	KITAMURA et al. 1982

nd, not determined

[a] Initiation within the *gag* leader sequence of avian retroviruses is indicated by "(*gag*)". In the cases of AEV-ES4, AEV-H, MC29, and MH2, it is not known whether translation of *erb*-B or *myc* is initiated in a *gag* leader sequence, or in the acquired cellular sequences. Termination within a viral gene, in a different reading frame from that used for the viral gene, is indicated by square brackets

[b] In several instances, the v-*onc* gene product terminates at the same position in each of two independently isolated viruses. It may be inferred that the cellular product has the same C-terminus

3 Oncogene Homologies

Based on their function and intracellular localization, the majority of oncogenes can be classified is four distinct groups (Table 1). Additional groups have been included which contain viral oncogenes whose function or subcellular location has not yet been determined, and non-viral oncogenes. Below we describe sequence homologies among the members of each group, and between the groups. For the figures shown in this review, we have outlined only identities between amino acids. In some regions, it is the charged or hydrophobic environment, rather than the specific residue, which is conserved (DAYHOFF et al. 1978). Though possibly significant, these functional homologies have not generally been included in the current discussion.

3.1 Tyrosine Protein Kinases

Despite an extensive knowledge of the structure of viral and cellular oncogenes, little is known about their functions. In 1978, ERIKSON and colleagues observed that the transforming protein encoded by the *src* oncogene had protein kinase activity (COLLETT and ERIKSON 1978). This seminal finding was quickly confirmed in several laboratories, and extended by HUNTER and SEFTON to show that phosphorylation by the *src* kinase occurred on tyrosine residues (HUNTER and SEFTON 1980; SEFTON 1985). In the past few years, a number of oncogenes with tyrosine protein kinase (Tyr-PK) activity have been isolated (HUNTER 1984). They represent retroviruses which have originated from different species containing a variety of oncogenes. By hybridization analysis, most of these oncogenes do not appear to be related; however, they do share extensive nucleotide sequence homology. Furthermore, the Tyr-PKs encoded by their oncogenes share even greater homology. Figure 2 shows a dot matrix representation of amino acid sequence homologies among viral Tyr-PK gene products, and Fig. 3 gives a comparison of the conserved regions (indicated by the *dotted lines* in Fig. 2) in each protein. Within the avian oncogenic proteins, *src* and *yes* (KITAMURA et al. 1982) are the most closely related. Other transforming proteins in this class are less homologous. The homologous regions of these gene products lie predominantly in their C-terminal regions, whereas the related portion of the *abl* protein lies near the N-terminal region of the acquired sequences (REDDY et al. 1983b; R. LEE, M. PASKIND, J.Y.J. WANG, and D. BALTIMORE 1984, personal communication). The Tyr-PK activity associated with these genes has been localized to the same homologous region. It thus appears that conservation of the Tyr-PK domain allows otherwise distinct proteins to manifest the same function.

3.2 Oncogenes Without Manifest Tyr-PK Activity

A large majority of retroviral oncogene products do not exhibit protein kinase activity. Many of them, however, do share some sequence homology with oncogenes encoding Tyr-PKs. One of the first examples of this type of homology was uncovered by comparing the p60^{src} sequence with the predicted 37-kD protein encoded by the *mos* gene of Moloney murine sarcoma virus (Mo-MSV) (VAN BEVEREN et al. 1981a). The C-terminal regions of *src*- and *mos*-encoded proteins exhibited over 25% sequence identity. Figure 3 displays the homologies among several oncogene products which share a degree of homology with p60$^{v\text{-}src}$. Regions homologous to the Tyr-PK proteins were apparent in the

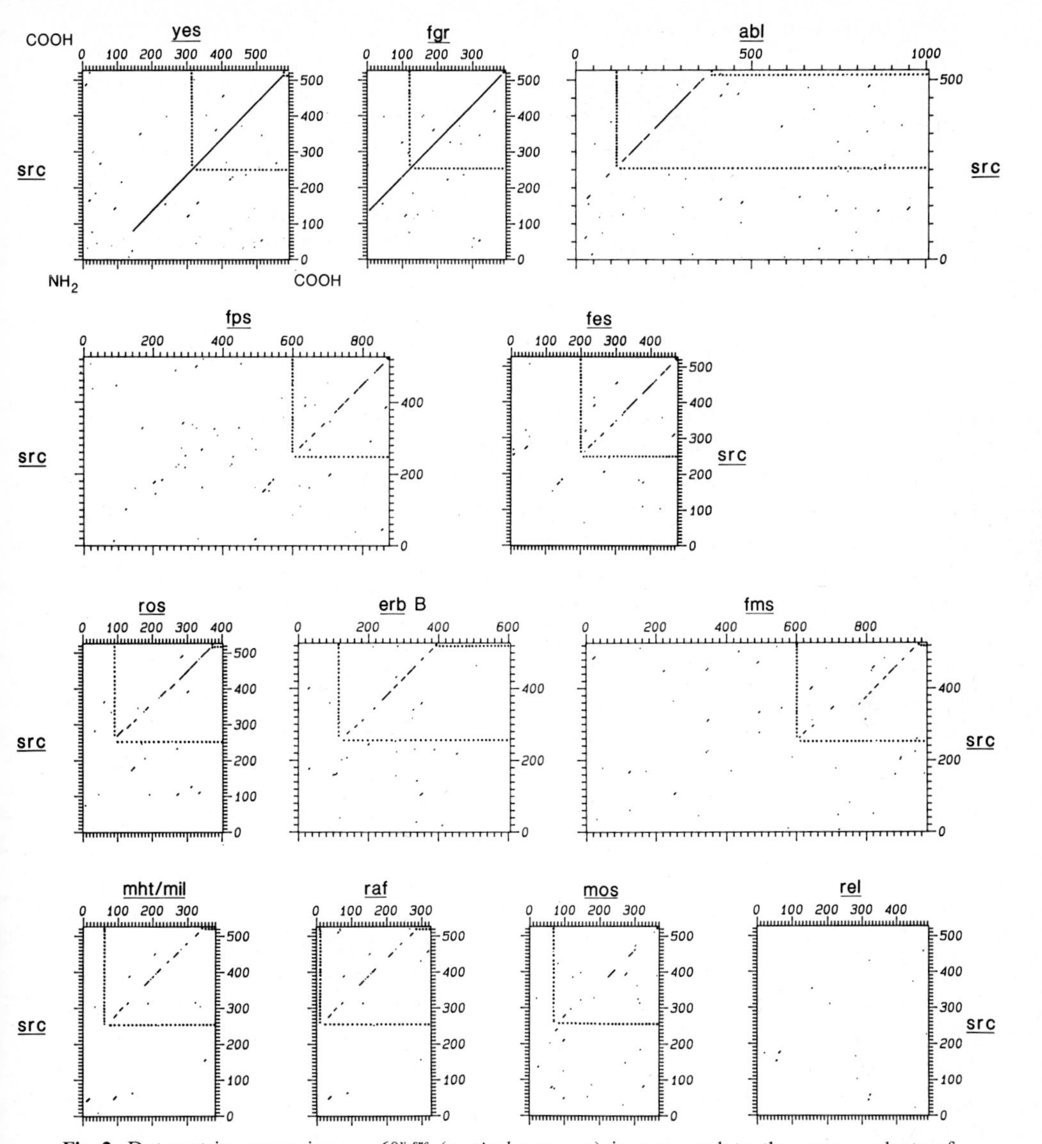

Fig. 2. Dot matrix comparisons. p60$^{v\text{-}src}$ (*vertical sequence*) is compared to the gene products of other members of the *src* family (*horizontal sequences*). *Position 1* in the *src* sequence indicates the first amino acid of p60$^{v\text{-}src}$. *Position 1* in each of the other sequences indicates the first residue in the v-*onc* portion of each gene product. The N-terminus of each sequence is at the *bottom left* portion of each plot, and the C-terminus at the *top* (*src*) or *right* (other *onc*) portions of the plots. A *dot* indicates identity at 5 of 11 amino acids in the two sequences being analyzed. A diagonal line can be discerned in regions of extended identity. The protein sequences were compared using the COMPARE and DOTPLOT dot matrix programs (MAIZEL and LENK 1981; DEVEREUX et al. 1984). Plots were printed at 100 dots per centimeter on a Hewlett Packard series 7221 terminal printer. Regions indicated by *boxes* are those shown in Fig. 3. The specific sequences used here are those for the products of Pr-RSV-C *src* (SCHWARTZ et al. 1983); Y73 *yes* (KITAMURA et al. 1982); GR-FeSV *fgr* (NAHARRO et al. 1984); Ab-MLV(P160) *abl* (R. LEE, M. PASKIND, J.Y.J. WANG, D. BALTIMORE 1984, personal communication); Fujinami sarcoma virus *fps* (SHIBUYA and HANAFUSA 1982); ST-FeSV *fes* (HAMPE et al. 1982); UR2 *ros* (NECKAMEYER and WANG 1985); AEV-H *erb*-B (YAMAMOTO et al. 1983); SM-FeSV *fms* (HAMPE et al. 1984); MH2 *mht*/*mil* (KAN et al. 1984b); 3611-MSV *raf* (MARK and RAPP 1984); Mo-MSV *mos* (VAN BEVEREN et al. 1981a); and REV-T *rel* genes (STEPHENS et al. 1983)

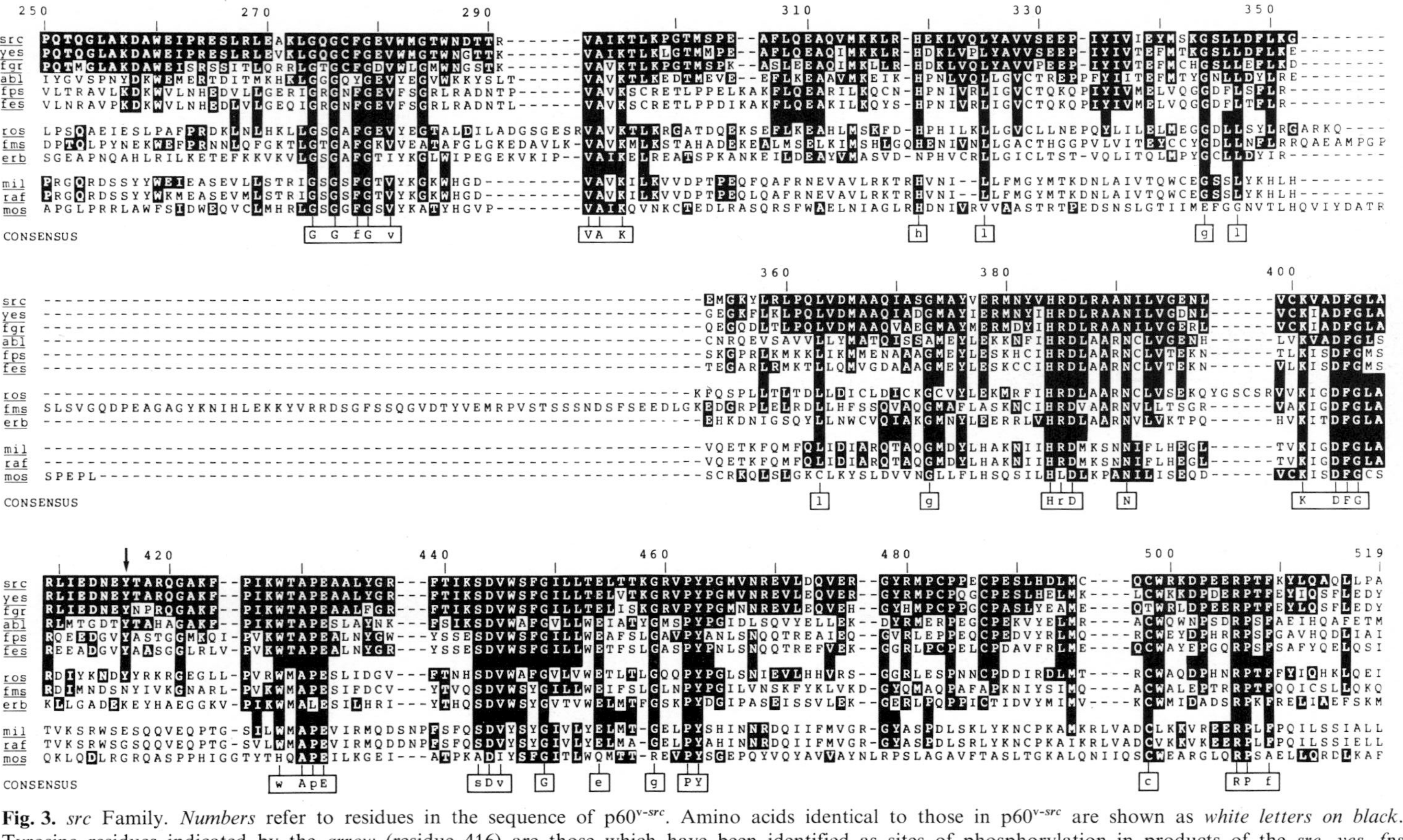

Fig. 3. *src* Family. *Numbers* refer to residues in the sequence of p60$^{v\text{-}src}$. Amino acids identical to those in p60$^{v\text{-}src}$ are shown as *white letters on black*. Tyrosine residues indicated by the *arrow* (residue 416) are those which have been identified as sites of phosphorylation in products of the *src*, *yes*, *fps* (PRCII), *fes* (ST-FeSV), and *abl* genes (PATSCHINSKY et al. 1982). Amino acids identical in all 12 sequences are indicated in *upper case letters* below the aligned sequences; those that are identical in 11 of the 12 sequences are indicated in *lower case letters*. Consecutive identical residues are indicated in *boxes*. To obtain the indicated alignment, the optimum alignment of v-*src* and v-*fms* gene products was first obtained with the ALIGN program using a unitary matrix and a gap penalty of 3 (DAYHOFF 1976). Other gene products were then added, with further gaps being introduced where necessary in order to optimize the alignment in the entire group. Sequences used are those referenced in Fig. 2

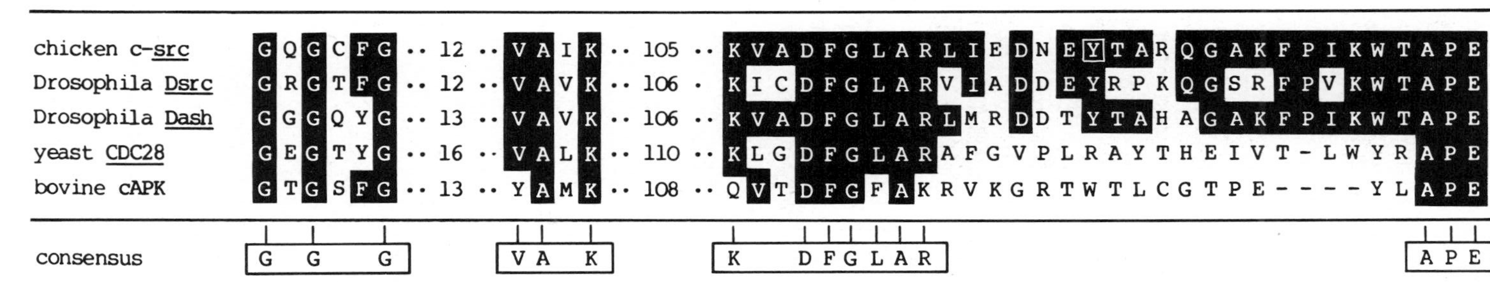

Fig. 4. Kinase conservation among species. Amino acids identical to those in p60$^{v\text{-}src}$ are shown as *white letters on black*. The tyrosine residue phosphorylated in p60$^{v\text{-}src}$ is indicated with a *white box*. Consensus sequences of residues are indicated in *boxes*. In addition to that of the Pr-RSV-C *src* gene, the sequences shown include the products of the *Drosophila Dsrc* and *Dash* genes (Hoffmann et al. 1983), the *Saccharomyces cerevisiae CDC28* gene (Lörincz and Reed 1984), and bovine cAMP-dependent protein kinase (Shoji et al. 1981). (A more extensive representation of this comparison may be seen in Hunter and Cooper 1985)

mht/mil, *raf*, and *mos* transforming proteins. As shown in Fig. 4, homologies were also noted when comparisons were made between *src* and bovine cyclic adenosine monophosphate (cAMP) kinase, and among the avian and *Drosophila src* proteins (BARKER and DAYHOFF 1982; HOFFMANN et al. 1983; SIMON et al. 1983). Furthermore, homology was demonstrated between a yeast cell division mutant, *CDC28*, and the *src* family kinase domain (LÖRINCZ and REED 1984).

Certain portions of the Tyr-PK domain are more conserved than are other regions of the proteins. The comparisons shown in Figs. 3 and 4 indicate striking conservation in certain portions of the protein. Furthermore, the relative position of some amino acids is constant. Such homologies are reminiscent of evolution from a common ancestor, and suggest that these homologous regions are perhaps essential for Tyr-PK activity and, consequently, for transformation. The tyrosine residue phosphorylated in most Class I proteins lies within this region (*arrow* in Fig. 3). Mutation of the p60$^{v\text{-}src}$ tyrosine-416 to phenylalanine does not affect the in vitro protein kinase activity of the resultant product, nor the ability of the virus to transform fibroblasts. However, mouse cells transformed with the mutated DNA are unable to cause tumors in syngeneic mice (SNYDER and BISHOP 1984). Alteration of the lysine residue in the Val-Ala-X-Lys region (VA-K; see Fig. 3) of the oncogenes of either RSV or Mo-MSV leads to the loss of their transforming ability (KAMPS et al. 1984; M. HANNINK and D. DONOGHUE 1984, personal communication). Additionally, alteration of the residues in the Ala-Pro-Glu sequence (APE: see ApE Fig. 3) abolishes transformation by RSV (Bryant and Parsons 1984). Similarly, other changes in this region also affect the tumor potential of transformed cells (CROSS and HANAFUSA 1983). It may be that the common regions shown in Fig. 3 identify portions of these oncogene products which form their active sites.

3.3 ras Oncogenes

Harvey sarcoma virus (Ha-MSV) and Kirsten sarcoma virus (Ki-MSV), which were isolated from rat tumors induced by MLV (murine leukemia virus), contain distinct *ras* oncogenes, which do not appear from hybridization analysis to be closely related (HARVEY 1964; KIRSTEN and MAYER 1967; ELLIS et al. 1981). Though two distinct loci in the cell from which *ras* oncogenes were acquired have been identified, the proteins encoded by the Ha-*ras* or Ki-*ras* oncogenes share extensive homologies (YOUNG et al. 1979; DHAR et al. 1982; ELLIS et al. 1982; TSUCHIDA et al. 1982; SHIMIZU et al. 1983a). The Ha-*ras* gene has also been acquired, from rat and mouse, in other genomic organizations. In Rasheed rat sarcoma virus, the rat Ha-*ras* gene is fused to the viral *gag* gene, while in BALB/c mouse sarcoma virus (BALB-MSV) the mouse homolog of the Ha-*ras* gene (termed *bas*) lies between the end of the *pol* gene and the middle of the *env* gene (ANDERSEN et al. 1981; YOUNG et al. 1981; RASHEED et al. 1983; REDDY et al. 1985).

The *ras* oncogenes are of particular interest, since 15%–20% of human tumors contain an altered form of the cellular homologs of Ha-*ras* and Ki-*ras* genes (COOPER 1982; SHIMIZU et al. 1983b; WEINBERG 1983). Additionally, a third *ras* gene, N-*ras*, has been identified as the transforming gene in certain neuroblastoma cell lines (SHIMIZU et al. 1983b; TAPAROWSKY et al. 1983). It appears that in many human tumors the c-Ha-*ras*, c-Ki-*ras*, or N-*ras* is activated by single-base-pair changes in the structural gene, encoding the 21000-dalton *ras* protein (TABIN et al. 1982; REDDY et al. 1982; TAPAROWSKY et al. 1982; YUASA et al. 1983; BOS et al. 1984). Alterations at positions 12 or 61 have been directly linked to induction of the transformed phenotype. In vitro mutagenesis studies show that, with the exception of glycine and proline residues, the insertion of any other amino acid at position 12 leads to cellular transformation (MCGRATH et al. 1984; SEEBURG et al. 1984).

Figure 5 shows that the *ras* protein is evolutionarily conserved, from yeast to man. The mammalian, *Drosophila*, and *Dictyostelium ras* proteins are all approximately the same length (185–189 residues), whereas the yeast *ras* proteins are larger in size (219–323 residues: DEFEO-JONES et al. 1983; GALLWITZ et al. 1983; DHAR et al. 1984; NEUMAN-SILBERBERG et al. 1984; POWERS et al. 1984; REYMOND et al. 1984; FUKUI and KAZIRO 1985). The yeast *ras* proteins are nearly 90% homologous to the first 80 positions of the mammalian *ras* proteins, and show nearly 50% homology to the next 80 amino acids. The two yeast *ras* genes, *RAS1* and *RAS2*, are more homologous to each other, with about 90% homology in the first 180 positions (KATAOKA et al. 1985). A sequence Gly-X-Gly-X-X-Gly, shared between members of the *src* and *ras* families (compare Figs. 3, 4, 5) has been found in other proteins sharing a nucleotide-binding domain (GAY and WALKER 1983; WIERENGA and HOL 1983). Amino acid 12, involved in the activation of the mammalian *ras* proteins, lies in this region.

The *ras* protein has been localized to the inner surface of the plasma membrane, and mutations which prevent this association interfere with transformation by an activated *ras* gene (WILLINGHAM et al. 1980; WILLUMSEN et al. 1984a, b). Both normal and mutant *ras* proteins bind GTP (guanosine triphosphate; SCOLNICK et al. 1979), and recently the normal product has beeen shown to manifest GTPase activity in vitro (GIBBS et al. 1984; MCGRATH et al. 1984; SWEET et al. 1984). Mutant *ras* proteins appear to be devoid of GTPase activity. Yeast *ras* genes can also complement adenyl cyclase mutants in yeast, suggesting a role in normal cells (TODA et al. 1985).

3.4 Oncogenes Whose Products Are Located in the Nucleus

In the absence of any attributable function of most oncogene products, their location in the cell might offer clues to their role. The *myc*, *myb*, *fos*, and *ski* oncogene products have been shown to be localized in the nucleus (ABRAMS et al. 1982; DONNER et al. 1982; ALITALO et al. 1983; HANN et al. 1983; BOYLE et al. 1984; CURRAN et al. 1984; KLEMPNAUER et al. 1984; D. BRODEUR, A. BARKAS, E. STAVNEZER, personal communication). Despite their nuclear location, neither the *myc*, *myb*, *fos*, or *ski* oncogenes, nor their products, share sequence homology to the degree exhibited by members of the Tyr-PK or *ras* families. It is possible that the nuclear proteins share a short sequence which

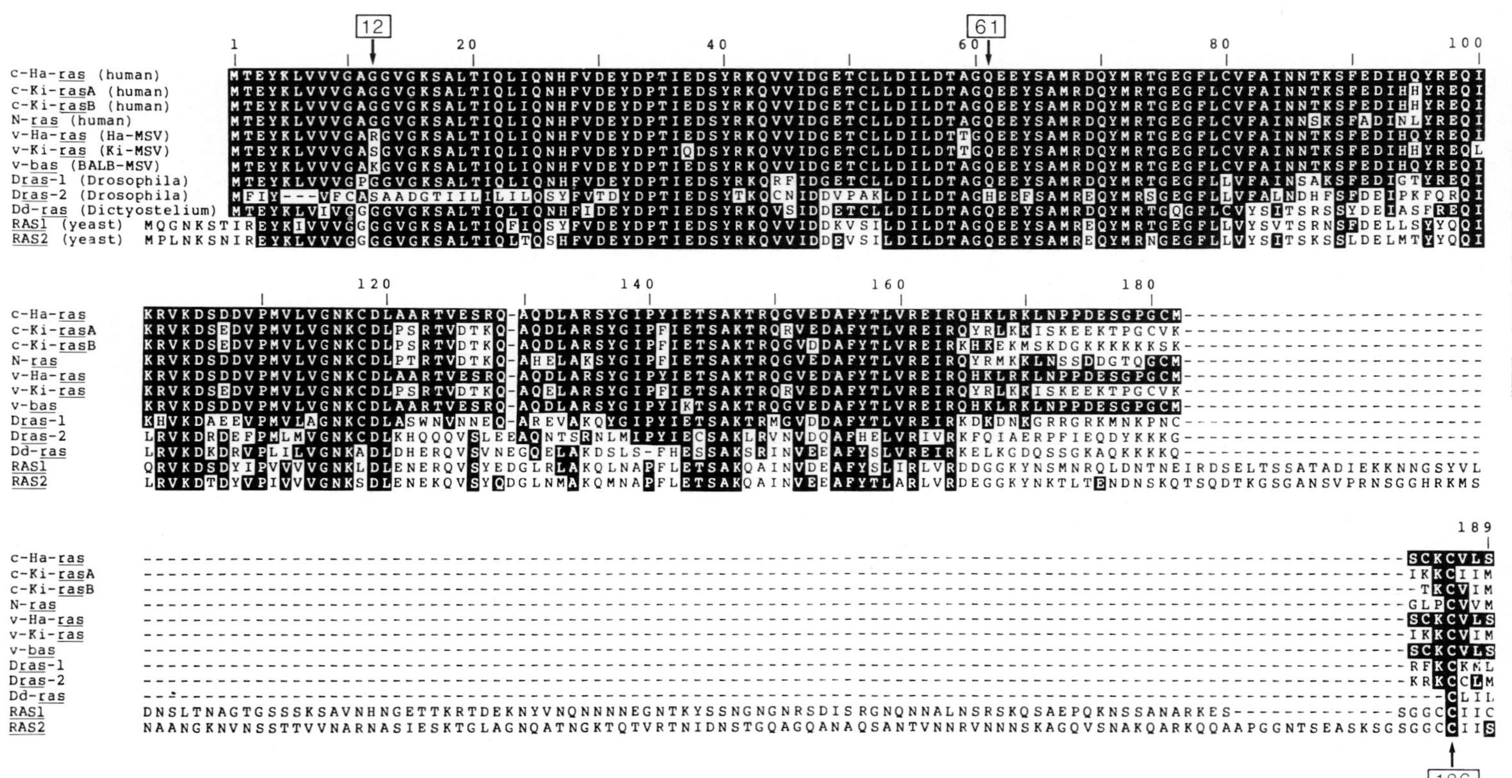

Fig. 5. *ras* Genes family. *Numbers* refer to residues in the sequence of human p21$^{c\text{-}Ha\text{-}ras}$. *Arrows* indicate residues 12 and 61, at which sequence alterations lead to activation of the normal gene, and residue 186, at which sequence alteration leads to loss of membrane association and transformation by the product of the Ha-MSV v-*ras*-transforming allele (Willumsen et al. 1984a, 1984b). Amino acids identical to those in the normal human p21$^{c\text{-}Ha\text{-}ras}$ (Capon et al. 1983) are shown as *white letters on black*. The sequences shown include those of products of the normal human c-Ki-*ras* gene (McGrath et al. 1983); the normal human N-*ras* gene (Taparowsky et al. 1983); the transforming, viral version of the rat Ha-*ras* gene ("v-Ha-*ras*"; Yasuda et al. 1984); the transforming, viral version of the rat Ki-*ras* gene ("v-Ki-*ras*"; Tsuchida et al. 1982); the transforming, viral version of the mouse Ha-*ras* gene ("v-*bas*"; Reddy et al. 1985); the (normal) *Drosophila Dras*-1 and *Dras*-2 genes (Neuman-Silberberg et al. 1984); the (normal) *Dictyostelium* Dt-*ras* gene (Reymond et al. 1984); and the (normal) *S. cerevisiae RAS1* and *RAS2* genes (Powers et al. 1984). The location of gaps is taken from Powers et al. (1984)

directs them to the nucleus, but such a putative nuclear signal sequence has thus far not been identified.

Two c-*myc*-related genes, referred to as N-*myc* and L-*myc*, have also been identified by hybridization with *myc*-specific sequences (KOHL et al. 1983; SCHWAB et al. 1983; J. MINNA 1984, personal communication). N-*myc* shares only limited sequence homology with the human c-*myc* gene, while the sequence of the L-*myc* gene is not yet known (SCHWAB et al. 1983; MICHITSCH and MELERA 1985). Similarly, an r-*fos* gene related to the mouse c-*fos* gene has also been identified (COCHRAN et al. 1984). Homology is in the third exon of the c-*fos* gene.

A nuclear location for a gene product is suggestive of a role in the regulation of gene expression. Indeed, one of the earliest changes observed upon stimulation of quiescent cells with mitogens is the rapid induction of both the c-*myc* and c-*fos* genes (KELLY et al. 1983; ARMELIN et al. 1984; COCHRAN et al. 1984; GREENBERG and ZIFF 1984; KRUIJER et al. 1984; MÜLLER et al. 1984). The kinetics of induction of c-*fos* mRNA shows a rapid but transient increase, with maximal accumulation of mRNA occurring 20–30 min postinduction. The c-*myc* mRNA attains a maximal level by 2–4 h, while that of r-*fos* is somewhat intermediate. The c-*fos* gene is also transiently expressed following partial hepatectomy, stimulation of T- and B-cells, or differentiation of the PC-12 cell line to neurites upon addition of nerve growth factor (NGF) or AMP (W. KRUIJER and D. SCHUBERT, personal communication). When promonocytes differentiate to macrophages, c-*fos* is also rapidly expressed, and the c-*fos* protein is transiently synthesized (MITCHELL et al. 1985). In HL-60 cells, c-*fos* expression is witnessed only when cells differentiate to macrophages, and not when the myelocytic cells differentiate to granulocytes. Expression of the c-*myc* and c-*myb* genes declines during differentiation. A role for the c-*myc* gene has been hypothesized in the cell cycle, but recent results indicate invariant levels of c-*myc* during the cell cycle (HANN et al. 1985; THOMPSON et al. 1985).

Another common feature of many nuclear oncogenic proteins is their ability to collaborate with other oncogenes to transform primary fibroblasts. These include *myc*, *myb*, *ski*, p53, adenovirus E1a protein, and SV40 and polyoma large T-antigens, which, along with the mutated form of the *ras* gene, can induce transformation of rat primary embryo fibroblasts (LAND et al. 1983; RULEY 1983). Many of the nuclear oncogenes can immortalize primary embryo fibroblasts.

3.5 Oncogenes Homologous to Growth Factors and Receptors

Since an archetypical hallmark of transformed cells is the loss of growth control, it has often been suggested that the products of oncogenes may act as mitogens or growth promoters. With the availability of nucleotide and amino acid sequences of a variety of oncogenes and bona fide growth factors and receptors, it has been feasible to determine structural relationships. First rewards of such endeavors came from the observation that the product of the *sis* oncogene was related to one of the two chains of platelet-derived growth factor (PDGF)

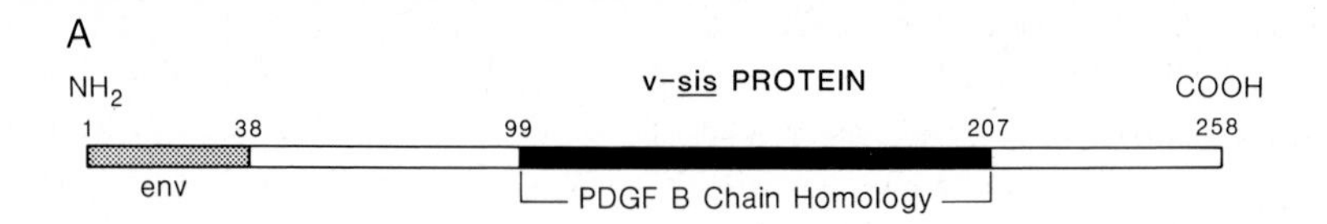

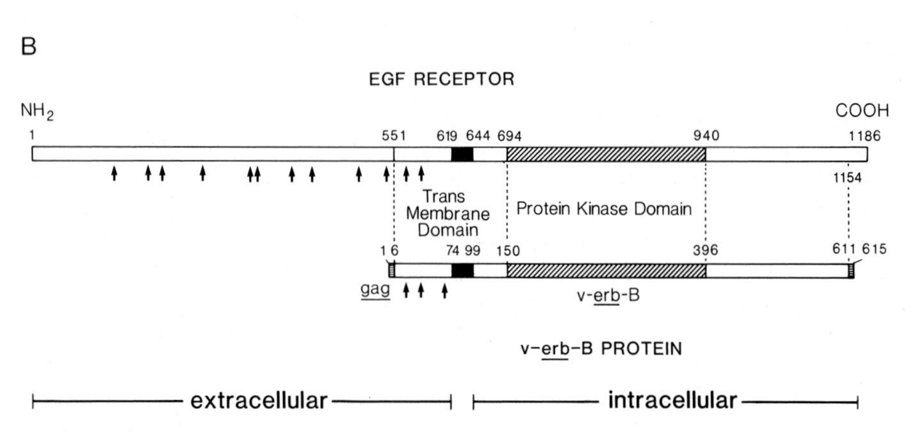

Fig. 6 A, B. Relationships between oncogene products, a growth factor, and a receptor. **A** Relationship of v-*sis* and human PDGF B-chain. The predicted structure of the v-*sis* gene product is based on the sequence of simian sarcoma virus and the initiation site of the v-*sis* protein (DEVARE et al. 1983; HANNINK and DONOGHUE 1984). The primary translation product would have 38 amino acids derived from the viral *env* gene, and 220 amino acids from *sis* sequences. The N-terminus of the human PDGF B-chain corresponds to residue 99, and the C-terminus to residue 207, of the v-*sis* gene product. **B** Homology between v-*erb*-B and human EGF receptor. The primary structure of the human EGF receptor derived from the nucleotide sequence is shown (ULLRICH et al. 1984). By comparison to the EGF receptor, the viral product lacks the N-terminal 579 and C-terminal 32 amino acids, and has homology to the receptor over 605 residues. *Vertical arrows:* glycosylation sites; *closed box:* transmembrane domain

(DOOLITTLE et al. 1983; WATERFIELD et al. 1983). Presumably, portions of the cellular gene encoding PDGF were acquired during biogenesis of simian sarcoma virus (SSV). Figure 6A outlines the relationship of the v-*sis* gene product with the B-chain of human PDGF. A region of over 100 residues (positions 99–207) shows differences at only three amino acids. The predicted v-*sis* gene product of 258 amino acids contains 38 amino acids of the *env* gene product at the N-terminus (DEVARE et al. 1983; JOHNSSON et al. 1984; HANNINK and DONOGHUE 1984). Homology begins at residue 99 and extends to residue 207 in the case of the human PDGF B-chain, while in the case of porcine PDGF the homology extends to residue 258 (CHIU et al. 1984; JOSEPHS et al. 1984; STROOBANT and WATERFIELD 1984). Although the complete structure of the c-*sis* gene at its 5′-end has not yet been completely deduced, there is little doubt that c-*sis* is the cellular gene encoding the B-chain of PDGF.

The molecular architecture of the v-*sis* gene might offer clues to its transforming ability. The v-*sis* gene was formed by insertion of c-*sis* sequences in the *env* gene of helper virus (simian sarcoma associated virus, SSAV) in such a way that the leader sequences are contributed by the viral *env* gene product (DEVARE et al. 1983; HANNINK and DONOGHUE 1984). The *env* gene product is a membrane protein, but the v-*sis* product is a secretory protein, because the *env* gene transmembrane sequences needed for anchorage in the membrane are deleted (DEUEL et al. 1983; OWEN et al. 1984; ROBBINS et al. 1983). Thus, the secreted v-*sis* protein in the medium can act as a growth factor in an autocrine fashion. In fact, there is evidence that medium from cells transformed by SSV is mitogenic, and that the activity can be blocked by anti-PDGF antibodies (DEUEL et al. 1983).

Another remarkable homology between an oncogene and a growth factor receptor was uncovered when the amino acid sequence of several tryptic peptides of the epidermal growth factor (EGF) receptor displayed a near-perfect fit with the predicted gene product of the v-*erb*-B oncogene of avian erythroblastosis virus (AEV) (DOWNWARD et al. 1984; LIN et al. 1984; XU et al. 1984). Figure 6B shows a diagrammatic representation of the extent and regions of homology between v-*erb*-B and the EGF receptor gene. The EGF receptor is 1186 amino acids long; roughly half of it lies inside the cell, while the other half constitutes the outer, EGF-binding domain. Residues 551–1154 of the human EGF receptor show over 90% identity with the chicken v-*erb*-B gene product (ULLRICH et al. 1984). Thus the v-*erb*-B gene product is a truncated form of the EGF receptor, lacking most of the external EGF-binding domain, but retaining the proposed membrane anchor domain. A 250-amino acid stretch of the homologous region has striking homology to the Tyr-PK domain (see Fig. 3). The v-*erb*-B gene product is glycosylated, and at least a fraction of it can be found on the surface of infected cells (HAYMAN and BEUG 1984; PRIVALSKY et al. 1984).

The structure of a number of other oncogenes bears an architectural resemblance to receptors. An immediate candidate is the product of the recently characterized *neu* gene, the oncogene detected in rat neuroglioblastomas, which encodes a 185-kD surface glycoprotein with an associated protein kinase activity (PADHY et al. 1982; SCHECHTER et al. 1984). The precise nature of the *neu* gene product remains unknown, but it appears to weakly hybridize to the v-*erb*-B gene. Another oncogene which has all the hallmarks of a receptor protein is the product of c-*fms*, whose viral cognate is the transforming gene of the McDonough strain of feline sarcoma virus (SM-FeSV) (HAMPE et al. 1984; RETTEMIER et al. 1985a). Nucleotide sequence analysis reveals that the v-*fms* and c-*fms* glycosylated proteins (120–170 kD in size) have a structure reminiscent of the EGF receptor gene: an external ligand-binding domain, a transmembrane anchor sequence, and an internal, cytoplasmic domain which has structural homologies to the Tyr-PK domain. Analysis of the v-*fms* product indicates that transported to the cell surface, with the ligand-binding domain external to the cell, and that the protein can be detected in clathrin-coated surface vesicles, as expected for a growth factor receptor (MANGER et al. 1984; RETTENMIER et al. 1985b). It may be that most of the viral Class I and IA Tyr-PK are truncated receptors.

The last "group" of oncogenes, *erb*-A, *ets*, and *rel* (Table 1), do not appear to belong to any of the other groups, largely owing to lack of information about their functions and cellular loci.

4 Comparison of c-onc and v-onc Products

The viral oncogenes have been acquired from normal cells. Do the acquired sequences undergo any change with respect to the cellular progenitors? In no case is the complete transcriptional unit of a proto-oncogene acquired. Thus all v-*onc*s have, in a sense, undergone a change compared to their cellular cognates. It can be argued that the inability of c-*onc* genes to induce cellular transformation, as compared to their viral counterparts, is a consequence of the alterations that the acquired sequences have undergone. A comparison of a number of viral and cellular oncogene products indicates that many v-*onc* gene products have undergone substantial changes. There are two main types of alterations.

1. Single-Base Differences. There are a number of single-base-pair changes, some of which result in changes in amino acid. It is difficult to assess the effect of these single-nucleotide changes on the biological activity by comparing the sequence. One way to understand the biological significance of single-base change is to construct recombinants containing portions of the altered and normal gene fragments, and study their ability to induce transformation. Such experiments revealed that cellular *ras* oncogenes can be biologically activated by single-nucleotide changes. For instance, a single-base change resulting in an altered amino acid at position 12 (glycine to valine) allows the cellular *ras* gene in human bladder carcinoma to manifest transforming activity (Tabin et al. 1982; Reddy et al. 1982). When the v-*mos* gene sequence was compared with its cellular homolog, c-*mos*, 25 single-base-pair changes were observed, of which 11 alter the corresponding amino acids (Van Beveren et al. 1981 b). The viruses containing the *mos* oncogene have been extensively propagated in the laboratory and may have undergone substantial mutations. In fact, comparisons of a number of strains of acutely transforming MSVs do not show alteration of the same bases. Thus the significance of single-base changes cannot be assessed in the absence of any biological data.

2. Frameshift Alterations. There are a number of cases in which substantial alteration has occurred in the coding domains of viral oncogenes as compared to their cellular progenitors. A striking example of the altered structure was uncovered when the v-*fos* and c-*fos* gene products were compared (Van Beveren et al. 1983). The c-*fos* gene in its fourth exon contains 104 additional nucleotides which were deleted during the biogenesis of the v-*fos* gene. Thus the reading frame at the site of deletion is altered in the two proteins. The alteration in the C-termini, however, does not appear to influence the ability of either v-*fos* or c-*fos* protein to induce cellular transformation (Miller et al. 1984).

Nucleotide sequence analysis of another *fos*-containing retrovirus, FBR murine osteosarcoma virus (FBR-MSV), which was isolated from a radiation-induced bone tumor, indicates that it synthesizes *gag-fos* fusion protein (Curran and Verma 1984; Van Beveren et al. 1984). The *fos* portion of the fusion protein is missing 24 amino acids at its N-terminus and 98 amino acids at its C-terminus, as compared to the c-*fos* protein. Additionally it has undergone two deletions and an addition of non-*fos* sequences at its 3′ end. Presumably, during the biogenesis of FBR-MSV, the *fos* gene recombined with another cellular gene, referred to as c-*fox*.

By comparison to their c-*onc* progenitors, several other v-*onc* products have altered C-termini (see Table 2). The v-*erb*-B gene product is about 30 amino acids shorter at its C-terminus than the EGF receptor (the product of the putative c-*erb*-B proto-oncogene). Similarly, the v-*fms* and c-*fms* proteins have different C-termini (C. Van Beveren, unpublished results). Alteration at the C-termini of the viral *myb* and *src* gene products has also been reported (Klempnauer et al. 1982, 1983; Takeya and Hanafusa 1983). The nine amino acids at the C-terminus of the 45-kD v-*myb* product differ from the 12 terminal residues of the 76-kD c-*myb* protein. It is not yet known if the alterations at the 3′ end of v-*myb* have biological significance. On the other hand, the alteration in the C-termini of v-*src* protein has a profound biological effect on cellular transformation. The C-terminal 12 amino acids of the $p60^{v\text{-}src}$ protein differ from the 18 residues at the C-terminus of c-*src*. The terminal 12 amino acids of v-*src* originate from a sequence 3′ to the c-*src* gene. Transformation of rodent cells by c-*src* can be observed if its C-terminal sequences are replaced by those of v-*src* and its adjacent LTR sequences (Iba et al. 1984; Parker et al. 1984; Shalloway et al. 1984). It is not obvious whether mere replacement of the 3′ end is sufficient to induce transformation by c-*src*, since chicken cells are not transformed by such constructs.

Many viral oncogenes form fused gene products which terminate either in the acquired cellular sequences or in a viral structural gene. Table 2 catalogs the nature of the N- or C-terminal region of the v-*onc* gene product in relation to their cellular progenitors. With few exceptions, most v-*onc* gene products are not identical to the products of their cellular cognates.

5 Structural Homologies

We have restricted this review to sequence identities among oncogenes. However, using other criteria, such as conservation of spacing of cysteine residues and charged regions, other, structural relationships have been discerned. In one study, the *myb* and *myc* genes were shown to be related to the human adenovirus-12 E1a gene product, a protein which, like the *myb* and *myc* products, appears in the nucleus (Ralston and Bishop 1983). The homology observed between *myc, myb,* and Ela could not be detected by dot matrix comparisons such as those shown in Fig. 2. There have also been reports of distant relationships between products of *rel* and the Tyr-PK family of genes (Stephens et al. 1983); *erb*-A and carbonic anhydrase (Debuire et al. 1984); *ets* and the yeast cell cycle genes *CDC4* and *CDC36* (Peterson et al. 1984); and the *Blym* gene and transferrin (Goubin et al. 1983).

6 Prospects of New Oncogenes

Are new oncogenes going to be discovered in the next few years, or have we reached the saturation limit? An answer to this query can be given more emphatically if one is permitted a caveat! Given the present technology of identification of transforming genes by DNA transfection, it is fair to assume that we will not witness the isolation of many new oncogenes. This position stems from two observations:

1. The same oncogene has been acquired by a number of different retroviruses. For instance, MC29, MH2, CMII, and OK10 avian retroviruses all contain the *myc* gene (BISTER et al. 1979, 1980). AMV and E26 contain the same oncogene, *myb* (KLEMPNAUER et al. 1982; NUNN et al. 1983). Fujinami sarcoma virus, PRCII, PRCIV, UR1, and 16L all contain the *fps* gene (SHIBUYA et al. 1980; BREITMAN et al. 1981; WANG et al. 1981; NEEL et al. 1982). The *fos* gene is present in the genomes of both FBJ-MSV and FBR-MSV (VAN BEVEREN et al. 1983, 1984).
2. Viruses isolated in one species often contain oncogenes previously isolated from other species. The *bas* oncogene of BALB-MSV is the mouse equivalent of the rat Ha-*ras* oncogene identified in both Ha-MSV and Rasheed rat sarcoma virus (REDDY et al. 1985). The *fes* gene identified in a number of feline sarcoma viruses is the homolog of the *fps* gene of FSV and PRCII viruses (SHIBUYA et al. 1980; HAMPE et al. 1982; SHIBUYA and HANAFUSA 1982). More recently it has been shown that the *mht/mil* oncogene of the MH2 virus is the avian homolog of the mouse *raf* oncogene of 3611-MSV (KAN et al. 1984a; MARK and RAPP 1984; SUTRAVE et al. 1984). Two recently isolated feline sarcoma viruses contain *abl* and *sis* genes, previously identified as oncogenes of the mouse virus Ab-MLV and the woolly monkey virus SSV, respectively (BESMER et al. 1983). Several isolates of viruses from thymic lymphosarcomas have contained the *myc* gene first identified in various avian viruses (LEVY et al. 1984; MULLINS 1984; NEIL et al. 1984).

Thus it appears that the possibility of finding new oncogenes is diminishing, owing, perhaps, to the limited probability that a given cellular sequence will recombine with a viral genome.

Though the majority of oncogenes were isolated in the genomes of acutely oncogenic retroviruses, there are additional oncogenes identified either by DNA transfection or hybridization that have not been found in a naturally occurring retrovirus. The *neu* gene, identified in DNA transfection studies from rat neuroglioblastoma DNA, has not been identified in any retrovirus (SCHECHTER et al. 1984). The B*lym* gene isolated from chicken bursal lymphomas or human Burkitt's lymphomas has, as yet, no counterpart in a retroviral genome (COOPER and NEIMAN 1980). Transforming genes have also been isolated and identified from human mammary tumors and a variety of T-cell lymphomas (COOPER 1982). Additionally, N-*ras* and N-*myc* genes have been identified by hybridization to *ras* and *myc* oncogenes, but no retrovirus containing N-*ras*, N-*myc* or B*lym* has yet been isolated.

Lack of identification of novel oncogenes may lie in the limitations of current techniques. The most successful transformation assays are performed on fibroblastic cell lines. Many oncogenes may not be scored positive for transformation in such cell lines, owing to lack either of expression or of the appropriate target substrate. With the identification of tissue-specific enhancers it may be possible to introduce *onc* genes which may express their transforming potential in specific cell types, for instance, lymphoid cells or epithelial cells. The finding that certain oncogenes realize their full transforming potential only when acting in concert may also aid in the identification of new oncogenes.

Availability of novel viral vector systems will also facilitate the identification of novel oncogenes. As more oncogenes are discovered, clearly the prospect of these genes being homologous to known cellular genes is very encouraging. It is easy to imagine that in the near future oncogenes will continue to dominate the field of cancer research.

Acknowledgements. The authors thank TONY HUNTER and BART SEFTON for their interest and helpful discussions, and CHRIS THURSTON for the design of the comparative figures. We are grateful to TONY HUNTER for Fig. 6. This work was supported by the National Cancer Institute and the American Cancer Society.

References

Abrams HD, Rohrschneider LR, Eisenman RN (1982) Nuclear location of the putative transforming protein of avian myelocytomatosis virus. Cell 29:427–439

Alitalo K, Ramsay G, Bishop JM, Pfeifer SO, Colby WW, Levinson AD (1983) Identification of nuclear proteins encoded by viral and cellular *myc* oncogenes. Nature 306:274–277

Andersen PR, Devare SG, Tronick SR, Ellis RW, Aaronson SA, Scolnick EM (1981) Generation of BALB-MuSV and Ha-MuSV by type C virus tranduction of homologous transforming genes from different species. Cell 26:129–134

Armelin HA, Armelin MCS, Kelly K, Stewart T, Leder P, Cochran BH, Stiles CD (1984) Functional role for c-*myc* in mitogenic response to platelet-derived growth factor. Nature 310:655–660

Barker WC, Dayhoff MO (1982) Viral *src* gene products are related to the catalytic chain of mammalian cAMP-dependent protein kinase. Proc Natl Acad Sci USA 79:2836–2839

Besmer P (1983) Acute transforming feline retroviruses. Curr Top Microbiol Immunol 107:1–27

Bishop JM (1983) Cellular oncogenes and retroviruses. Ann Rev Biochem 52:301–354

Bishop JM, Varmus HE (1982) Functions and origins of retroviral transforming genes. In: Weiss R et al. (eds) Molecular Biology of Tumor Viruses, 2nd edn. Cold Spring Harbor Laboratory, Cold Spring Harbor, New York, pp 999–1108

Bister K, Löliger HC, Duesberg PH (1979) Oligoribonucleotide map and protein of CMII: detection of conserved and nonconserved genetic elements in avian acute leukemia viruses CMII, MC29, and MH2. J Virol 32:208–219

Bister K, Ramsay G, Hayman MJ, Duesberg PH (1980) OK10, an avian acute leukemia virus of the MC29 subgroup with a unique genetic structure. Proc Natl Acad Sci USA 77:7142–7146

Bos JL, Verlaan-de Vries M, Jansen AM, Veeneman AH, van Boom JH, van der Eb AJ (1984) Three different mutations in codon 61 of the human N-*ras* gene detected by synthetic oligonucleotide hybridization. Nucl Acids Res 12:9155–9116

Boyle WJ, Lampert MA, Lipsick JS, Baluda MA (1984) Avian myeloblastosis virus and E26 virus oncogene products are nuclear proteins. Proc Natl Scad Sci USA 81:4265–4269

Breitman ML, Hirano A, Wong T, Vogt PK (1981) Characteristics of avian sarcoma virus PRCIV and comparison with PRCII-p. Virology 114:451–462

Bryant DL, Parsons JT (1984) Amino acid alterations within a highly conserved region of the Rous sarcoma virus *src* gene product pp60^{src} inactivate tyrosine protein kinase activity. Mol Cell Biol 4:862–866

Capon DJ, Chen EY, Levinson AD, Seeburg PH, Goeddel DV (1983) Complete nucleotide sequences of the T24 human bladder carcinoma oncogene and its normal homologue. Nature 302:33–37

Chiu I-M, Reddy EP, Givol D, Robbins KC, Tronick SR, Aaronson SA (1984) Nucleotide sequence analysis identifies the human c-*sis* proto-oncogene as a structural gene for platelet-derived growth factor. Cell 37:123–129

Cochran BH, Zullo J, Verma IM, Stiles CD (1984) Expression of the c-*fos* gene and of an *fos*-related gene is stimulated by platelet-derived growth factor. Science 226:1080–1082

Collett MS, Erikson RL (1978) Protein kinase activity associated with the avian sarcoma virus *src* gene product. Proc Natl Acad Sci USA 75:2021–2024

Cooper GM (1982) Cellular transforming genes. Science 217:801–806

Cooper GM, Neiman PE (1980) Transforming genes of neoplasms induced by avian lymphoid leukosis viruses. Nature 287:656–659

Cross FR, Hanafusa H (1983) Local mutagenesis of Rous sarcoma virus: the major sites of tyrosine and serine phosphorylation of $p60^{src}$ are dispensable for transformation. Cell 34:597–608

Curran T, Verma IM (1984) FBR murine osteosarcoma virus. I. Molecular analysis and characterization of a 75000-Da *gag-fos* fusion product. Virology 135:218–228

Czernilofsky AP, Levinson AD, Varmus HE, Bishop JM, Tischer E, Goodman H (1983) Corrections to the nucleotide sequence of the *src* gene of Rous sarcoma virus. Nature 301:736–738

Dayhoff MO (1976) Survey of new data and computer methods of analysis. In: Dayhoff MO (ed) Atlas of Protein Sequence and Structure, vol 5 suppl 2, National Biomedical Research Foundation, Washington DC, pp 1–8

Dayhoff MO, Schwartz RM, Orcutt BC (1978) A model of evolutionary change in proteins. In: Dayhoff MO (ed) Atlas of Protein Sequence and Structure, vol 5 suppl 3, National Biomedical Research Foundation, Washington DC, pp 345–352

Debuire B, Henry C, Benaissa M, Biserte G, Claverie JM, Saule S, Martin P, Stehelin D (1984) Sequencing the *erb*-A gene of avian erythroblastosis virus reveals a new type of oncogene. Science 224:1456–1459

DeFeo-Jones D, Scolnick EM, Koller R, Dhar R (1983) *ras*-Related gene sequences identified and isolated from *Saccharomyces cerevisiae*. Nature 306:707–709

Deuel TF, Huang JS, Huang SS, Stroobant P, Waterfield MD (1983) Expression of a platelet-derived growth factor-like protein in simian sarcoma virus transformed cells. Science 221:1348–1350

Devare SG, Reddy EP, Law JD, Robbins KC, Aaronson SA (1983) Nucleotide sequence of the simian sarcoma virus genome: demonstration that its acquired cellular sequences encode the transforming gene product $p28^{sis}$. Proc Natl Acad Sci USA 80:731–735

Devereux J, Haeberli P, Smithies O (1984) A comprehensive set of sequence analysis programs for the VAX. Nucl Acids Res 12:387–395

Dhar R, Ellis RW, Shih TY, Oroszlan S, Shapiro B, Maizel J, Lowy D, Scolnick E (1982) Nucleotide sequence of the p21 transforming protein of Harvey murine sarcoma virus. Science 217:934–937

Dhar R, Nieto A, Koller R, DeFeo-Jones D, Scolnick E (1984) Nucleotide sequence of two *ras*H related-genes isolated from the yeast *Saccharomyces cerevisiae*. Nucl Acids Res 12:3611–3618

Donner P, Greiser-Wilke I, Moelling K (1982) Nuclear localization and DNA binding of the transforming gene product of avian myelocytomatosis virus. Nature 296:262–266

Donoghue DJ (1982) Demonstration of biological activity and nucleotide sequence of an in vitro synthesized clone of the Moloney murine sarcoma virus *mos* gene. J Virol 42:538–546

Doolittle RF, Hunkapiller MW, Hood LE, Devare SG, Robbins KC, Aaronson SA, Antoniades HN (1983) Simian sarcoma virus *onc* gene, v-*sis*, is derived from the gene (or genes) encoding a platelet-derived growth factor. Science 221:275–277

Downward J, Yarden Y, Mayes E, Scrace G, Totty N, Stollwell P, Ullrich A, Schlessinger J, Waterfield MD (1984) Close similarity of epidermal growth factor receptor and v-*erb*-B oncogene protein sequences. Nature 307:521–527

Ellis RW, DeFeo D, Shih TY, Gonda MA, Young HA, Tsuchida N, Lowy DR, Scolnick EM (1981) The p21 *src* genes of Harvey and Kirsten sarcoma viruses originate from divergent members of a family of normal vertebrate genes. Nature 292:506–511

Ellis RW, DeFeo D, Furth ME, Scolnick EM (1982) Mouse cells contain two distinct *ras* gene mRNA species that can be translated into a p21 *onc* protein. Mol Cell Biol 2:1339–1345

Fukui Y, Kaziro Y (1975) Molecular cloning and sequence analysis of a *ras* gene from *Schizosaccharomyces pombe*. EMBO J 4:687–691

Galibert F, Dupont de Dinechin S, Righi M, Stehelin D (1984) The second oncogene *mil* of avian retrovirus MH2 is related to the *src* gene family. EMBO J 3:1333–1338

Gallwitz D, Donath C, Sander C (1983) A yeast gene encoding a protein homologous to the human c-*has/bas* proto-oncogene product. Nature 306:704–707

Gay NJ, Walker JE (1983) Homology between human bladder carcinoma oncogene product and mitochondrial ATP-synthase. Nature 301:262–264

Gibbs JB, Sigal IS, Poe M, Scolnick EM (1984) Intrinsic GTPase activity distinguishes normal and oncogenic *ras* p21 molecules. Proc Natl Acad Sci USA 81:5704–5708

Goubin G, Goldman DS, Luce J, Neiman PE, Cooper GM (1983) Molecular cloning and nucleotide sequence of a transforming gene detected by transfection of chicken B-cell lymphoma DNA. Nature 302:114–119

Greenberg ME, Ziff EB (1984) Stimulation of 3T3 cells induces transcription of the c-*fos* proto-oncogene. Nature 311:433–438

Hampe A, Laprevotte I, Galibert F, Fedele LA, Sherr CJ (1982) Nucleotide sequences of feline retroviral oncogenes (v-*fes*) provide evidence for a family of tyrosine-specific protein kinase genes. Cell 30:775–785

Hampe A, Gobet M, Scherr CJ, Galibert F (1984) Nucleotide sequence of the feline retroviral oncogene v-*fms* shows unexpected homology with oncogenes encoding tyrosine-specific protein kinases. Proc Natl Acad Sci USA 81:85–89

Hann SR, Abrams HR, Rohrschneider LR, Eisenman RN (1983) Proteins encoded by v-*myc* and c-*myc* oncogenes: identification and localization in acute leukemia virus transformants and bursal lymphoma cell lines. Cell 34:789–798

Hann SR, Thompson CB, Eisenman RN (1985) c-*myc* oncogene protein synthesis is independent of the cell cycle in human and avian cells. Nature 314:366–369

Hannink M, Donoghue DJ (1984) Requirement for a signal sequence in biological expression of the v-*sis* oncogene. Science 226:1197–1199

Harvey JJ (1964) An unidentified virus which causes the rapid production of tumours in mice. Nature 204:1104–1105

Hayman MJ, Beug H (1984) Identification of a form of the avian erythroblastosis virus *erb*-B gene product at the cell surface. Nature 309:460–462

Henry C, Coquillaud M, Saule S, Stehelin D, Debuire B (1985) The four C-terminal amino acids of the v-*erbA* polypeptide are encoded by an intronic sequence of the v-*erb*-B oncogene. Virology 140:179–182

Hoffmann FM, Fresco LD, Hoffman-Falk H, Shilo B-Z (1983) Nucleotide sequences of the Drosophila *src* and *abl* homologs: conservation and variability in the *src* family oncogenes. Cell 35:393–401

Huang C-C, Hammond C, Bishop JM (1984) Nucleotide sequence of v-*fps* in the PRCII strain of avian sarcoma virus. J Virol 50:125–131

Huang C-C, Hammond C, Bishop JM (1985) Nucleotide Sequence and Topography of Chicken c-*fps*.Genesis of a Retroviral Oncogene Encoding a Tyrosine-specific Protein Kinase. J Mol Biol 181:175–186

Hunter T (1984) The proteins of oncogenes. Scient Amer 251(2):70–79

Hunter T, Cooper JA (1985) Protein-tyrosine kinases. Ann Rev Biochem 54:897–930

Hunter T, Sefton BM (1980) Transforming gene product of Rous sarcoma virus phosphorylates tyrosine. Proc Natl Acad Sci USA 77:1311–1315

Iba H, Takeya T, Cross FR, Hanafusa T, Hanafusa H (1984) Rous sarcoma virus variants that carry the cellular *src* gene instead of the viral *src* gene cannot transform chicken embryo fibroblasts. Proc Natl Acad Sci USA 81:4424–4428

Johnsson A, Heldin C-H, Wasteson A, Westermark B, Deuel TF, Huang JS, Seeburg PH, Gray A, Ullrich A, Scarce G, Stroobant P, Waterfield MD (1984) The c-*sis* gene encodes a precursor of the B chain of platelet-derived growth factor. EMBO J 3:921–928

Josephs SF, Ratner L, Clarke MF, Westin EH, Reitz MS, Wong-Staal F (1984) Transforming potential of human c-*sis* nucleotide sequences encoding platelet-derived growth factor. Science 225:636–639

Kamps MP, Taylor SS, Setton BM (1984) Direct evidence that oncogenic tyrosine kinases and cyclic AMP-dependent protein kinase have homologous ATP-binding sites. Nature 310:589–592

Kan NC, Flordellis CS, Mark GE, Duesberg PH, Papas TS (1984a) A common *onc* gene sequence transduced by avian carcinoma virus MH2 and by murine sarcoma virus 3611. Science 223:813–816

Kan NC, Flordellis CS, Mark GE, Duesberg PH, Papas TS (1984b) Nucleotide sequence of avian carcinoma virus MH2: two potential *onc* genes, one related to avian virus MC29 and the other related to murine sarcoma virus 3611. Proc Natl Acad Sci USA 81:3000–3004

Kataoka T, Powers S, Cameron S, Fasano O, Goldfarb M, Broach J, Wigler M (1985) Functional homology of mammalian and yeast *RAS* genes. Cell 40:19–26

Kelly K, Cochran BH, Stiles CD, Leder P (1983) Cell-specific regulation of the c-*myc* gene by lymphocyte mitogens and platelet-derived growth factor. Cell 35:603–610

Kirsten WH, Mayer LA (1967) Morphologic responses to a murine erythroblastosis virus. J Natl Cancer Inst 39:311–335

Kitamura N, Kitamura A, Toyoshima K, Hirayama Y, Yoshida M (1982) Avain sarcoma virus Y73 genome sequence and structural similarity of its transforming gene product to that of Rous sarcoma virus. Nature 297:205–208

Klempnauer K-H, Gonda TJ, Bishop JM (1982) Nucleotide sequence of the retroviral leukemia gene v-*myb* and its cellular progenitor c-*myb*: the architecture of a transduced oncogene. Cell 31:453–463

Klempnauer K-H, Ramsay G, Bishop JM, Moscovici MG, Moscovici C, McGrath JP, Levinson AD (1983) The product of the retroviral transforming gene v-*myb* is a truncated version of the protein encoded by the cellular oncogene c-*myb*. Cell 33:345–355

Klempnauer K-H, Symonds G, Evan GI, Bishop JM (1984) Subcellular localization of proteins encoded by oncogenes of avian myeleoblastosis virus and avian leukemia virus E26 and by the chicken c-*myb* gene. Cell 37:537–547

Kohl NE, Kanda N, Schreck RR, Bruns G, Latt SA, Gilbert F, Alt FW (1983) Transposition and amplification of oncogene-related sequences in human neuroblastomas. Cell 35:359–367

Kruijer W, Cooper JA, Hunter T, Verma IM (1984) Platelet-derived growth factor induces rapid but transient expression of the c-*fos* gene and protein. Nature 312:711–716

Kruijer W, Schubert D, Verma IM (1985) Induction of the protooncogene *fos* by nerve growth factor. Proc Natl Acad Sci USA 82 (in press)

Land H, Parada LF, Weinberg RA (1983) Tumorigenic conversion of primary embryo fibroblasts requires at least two cooperating oncogenes. Nature 304:596–602

Levy LS, Gardner MB, Casey JW (1984) Isolation of a feline leukaemia provirus containing the oncogene *myc* from a feline lymphosarcoma. Nature 308:853–856

Lin CR, Chen WS, Kruiger W, Stolarsky LS, Weber W, Evans RM, Verma IM, Gill GN, Rosenfeld MG (1984) Expression cloning of human EGF receptor complementary DNA: gene amplification and three related messenger RNA products in A431 cells. Science 224:843–848

Lörincz AT, Reed SI (1984) Primary structure homology between the product of yeast cell division control gene *CDC28* and vertebrate oncogenes. Nature 307:183–185

Maizel JV Jr, Lenk RP (1981) Enhanced graphic matrix analysis of nucleic acid and protein sequences. Proc Natl Acad Sci USA 78:7665–7669

Manger R, Najita L, Nichols EJ, Hakomori S-I, Rohrschneider L (1984) Cell surface expression of the McDonough strain of feline sarcoma virus *fms* gene product ($gp140^{fms}$). Cell 39:327–337

Mann R, Mulligan RC, Baltimore D (1983) Construction of a retrovirus packaging mutant and its use to produce helper-free defective retrovirus. Cell 33:153–159

Mark GE, Rapp UR (1984) Primary structure of v-*raf*: relatedness to the *src* family of oncogenes. Science 224:285–289

McGrath JP, Capon DJ, Smith DH, Chen EY, Seegurg PH, Goeddel DV, Levinson AD (1983) Structure and organization of the human Ki-*ras* proto-oncogene and a related processed pseudogene. Nature 304:501–506

McGrath JP, Capon DJ, Goeddel DV, Levinson AD (1984) Comparative biochemical properties of normal and activated human *ras* p21 protein. Nature 310:644–649

Michitsch RW, Melera PW (1985) Nucleotide sequence of the 3′ exon of the human N-*myc* gene. Nucl Acids Res 13:2545–2558

Miller AD, Curran T, Verma IM (1984) c-*fos* protein can induce cellular transformation: a novel mechanism of activation of a cellular oncogene. Cell 36:51–60

Miller AD, Law M-F, Verma IM (1985) Generation of helper-free amphotropic retroviruses that transduce a dominant-acting, methotrexate-resistant dihydrofolate reductase gene. Mol Cell Biol 5:431–437

Mitchell RL, Zokas L, Schreiber RD, Verma IM (1985) Rapid induction of the expression of proto-oncogene *fos* during human monocytic differentiation. Cell 40:209–217

Müller R, Bravo R, Burckhardt J, Curran T (1984) Induction of c-*fos* gene and protein by growth factors precedes activation of c-*myc*. Nature 312:716–720

Müller R, Verma IM (1984) Expression of cellular oncogenes. Curr Topics Microbiol Immunol 112:73–115

Mullins JI, Brody DS, Binari RC Jr, Cotter SM (1984) Viral transduction of c-*myc* gene in naturally occurring feline leukaemias. Nature 308:856–858

Naharro G, Robbins KC, Reddy EP (1984) Gene product of v-*fgr onc*: hybrid protein containing a portion of actin and a tyrosine-specific protein kinase. Science 223:63–66

Neckameyer WS, Wang L-H (1985) Nucleotide sequence of avian sarcoma virus UR2 and comparison of its transforming gene with other members of the tyrosine protein kinase oncogene family. J Virol 53:879–884

Neel BG, Wang L-H, Mathey-Prevot B, Hanafusa T, Hanafusa H, Hayward WS (1982) Isolation of 16L virus: a rapidly transforming sarcoma virus from an avian leukosis virus-induced sarcoma. Proc Natl Acad Sci USA 79:5088–5092

Neil JC, Hughes D, McFarlane R, Wilkie NM, Onions DE, Lees G, Jarrett O (1984) Transduction and rearrangement of the *myc* gene by feline leukaemia virus in naturally occurring T-cell leukaemias. Nature 308:814–820

Neuman-Silberberg FS, Schejter E, Hoffmann FM, Shilo B-Z (1984) The Drosophila *ras* oncogenes: structure and nucleotide sequence. Cell 37:1027–1033

Nunn MF, Seeburg PH, Moscovici C, Duesberg PH (1983) Tripartite structure of the avian erythroblastosis virus E26 transforming gene. Nature 306:391–395

Owen AJ, Pantazis P, Antoniades HN (1984) Simian sarcoma virus-transformed cells secrete a mitogen identical to platelet-derived growth factor. Science 225:54–56

Padhy LC, Shih C, Cowing D, Finkelstein R, Weinberg RA (1982) Identification of a phosphoprotein specifically induced by the transforming DNA of rat neuroblastomas. Cell 28:865–871

Parker RC, Varmus HE, Bishop JM (1984) Expression of v-*src* and chicken c-*src* in rat cells demonstrates qualitative differences between pp60$^{v\text{-}src}$ and pp60$^{c\text{-}src}$. Cell 37:131–139

Patschinsky T, Hunter T, Esch FS, Cooper JA, Sefton BM (1982) Analysis of the sequence of amino acids surrounding sites of tyrosine phosphorylation. Proc Natl Acad Sci USA 79:973–977

Peterson TA, Yochem J, Byers B, Nunn MF, Duesberg PH, Doolittle RF, Reed SI (1984) A relationship between the yeast cell cycle genes *CDC4* and *CDC36* and the *ets* sequence of oncogenic virus E26. Nature 309:556–558

Powers S, Kataoka T, Fasano O, Goldfarb M Strathern J, Broach J, Wigler M (1984) Genes in S. cerevisiae encoding proteins with domains homologous to the mammalian *ras* proteins. Cell 36:607–612

Privalsky ML, Ralston R, Bishop JM (1984) The membrane glycoprotein encoded by the retroviral oncogene v-*erb*-B is structurally related to the tyrosine-specific protein kinase. Proc Natl Acad Sci USA 704–707

Ralston R, Bishop JM (1983) The protein products of the *myc* and *myb* oncogenes and adenovirus E1a are structurally related. Nature 306:803–806

Rasheed S, Norman GL, Heidecker G (1983) Nucleotide sequence of the Rasheed rat sarcoma virus oncogene: new mutations. Science 221:155–157

Reddy EP, Smith MJ, Aaronson SA (1981) Complete nucleotide sequence and organization of the Moloney murine sarcoma virus genome. Science 214:445–450

Reddy EP, Reynolds RK, Santos E, Barbacid M (1982) A point mutation is responsible for the acquisition of transforming properties by the T24 human bladder carcinoma oncogene. Nature 300:149–152

Reddy EP, Reynolds RK, Watson DK, Schultz RA, Lautenberger J, Papas TS (1983a) Nucleotide sequence analysis of the proviral genome of avian myelocytomatosis virus (MC29). Proc Natl Acad Sci USA 80:2500–2504

Reddy EP, Smith MJ, Srinivasan A (1983b) Necleotide sequence of Abelson murine leukemia virus genome: structural similarity of its transforming gene product to other *onc* gene products with tyrosine-specific kinase activity. Proc Natl Acad Sci USA 80:3623–3627

Reddy EP, Lipman D, Andersen PR, Tronick SR, Aaronson SA (1985) Nucleotide sequence analysis of the BALB/c murine sarcoma virus transforming gene. J Virol 53:984–987

Rettenmier CW, Chen JH, Roussel MF, Sherr CJ (1985a) The product of the c-*fms* proto-oncogene: a glycoprotein with associated tyrosine kinase activity. Science 228:320–322

Rettenmier CW, Roussel MF, Quinn CO, Kitchingman GR, Look AT, Sherr CJ (1985b) Transmembrane orientation of glycoproteins encoded by the v-*fms* oncogene. Cell 40:971–981

Reymond CD, Gomer RH, Mehdy MC, Firtel RA (1984) Developmental regulation of a Dictyostelium gene encoding a protein homolgous to mammalian *ras* protein. Cell 39:141–148

Robbins KC, Antoniades HN, Devare SG, Hunkapiller MW, Aaronson SA (1983) Structural and immunological similarities between simian sarcoma virus gene product(s) and human platelet-derived growth factor. Nature 305:605–608

Ruley HE (1983) Adenovirus early region 1A enables viral and cellular transforming genes to transform primary cells in culture. Nature 304:602–606

Rushlow KE, Lautenberger JA, Papas TS, Baluda MA, Perbal B, Chirikjian JG, Reddy EP (1982) Nucleotide sequence of the transforming gene of avian myeloblastosis virus. Science 216:1421–1423

Schechter AL, Stern DF, Vaidyanathan L, Decker SJ, Drebin JA, Greene MI, Weinberg RA (1984) The *neu* oncogene: an *erb*-B-related gene encoding a 185000-M_r tumour antigen. Nature 312:513–516

Schwab M, Alitalo K, Klempnauer K-H, Varmus HE, Bishop JM Gilbert F, Brodeur G, Goldstein M, Trent J (1983) Amplified DNA with limited homology to *myc* cellular oncogene is shared by human neuroblastoma cell lines and a neuroblastoma tumour. Nature 305:245–248

Schwartz DE, Tizard R, Gilbert W (1983) Nucleotide sequence of Rous sarcoma virus. Cell 32:853–869

Scolnick EM, Papageorge AG, Shih TY (1979) Guanine nucleotide-binding activity as an assay for *src* protein of rat-derived murine sarcoma viruses. Proc Natl Acad Sci USA 76:5355–5359

Seeburg PH, Colby WW, Capon DJ, Goeddel DV, Levinson AD (1984) Biological properties of human c-Ha-*ras*1 genes mutated at codon 12. Nature 312:71–75

Sefton BM (1985) The viral tyrosine protein kinases. Curr Topics Microbiol Immunol, this volume

Shalloway D, Coussens PM, Yaciuk P (1984) Overexpression of the s-*src* protein does not induce transformation of NIH 3T3 cells. Proc Natl Acad Sci USA 81:7071–7075

Shibuya M, Hanafusa T, Hanafusa H, Stephenson JR (1980) Homology exists among the transforming sequences of avian and feline sarcoma viruses. Proc Natl Acad Sci USA 77:6536–6540

Shibuya M, Hanafusa H (1982) Nucleotide sequence of Fujinami sarcoma virus: evolutionary relationship of its transforming gene with transforming genes of other sarcoma viruses. Cell 30:787–795

Shimizu K, Birnbaum D, Ruley MA, Fasano O, Suard Y, Edlund L, Taparowsky E, Goldfarb M, Wigler M (1983a) Structure of the Ki-*ras* gene of the human lung carcinoma cell line Calu-1. Nature 304:497–500

Shimizu K, Goldfarb M, Suard Y, Perucho M, Li Y, Kamata T, Feramisco J, Stavnezer E, Fogh J, Wigler MH (1983b) Three human transforming genes are related to the viral *ras* oncogenes. Proc Natl Acad Sci USA 80:2112–2116

Shoji S, Parmelee DC, Wade RD, Kumar S, Ericsson LH, Walsh KA, Neurath H, Long GL, Demaille JG, Fischer EH, Titani K (1981) Complete amino acid sequence of the catalytic subunit of bovine cardiac muscle cyclic AMP-dependent protein kinase. Proc Natl Acad Sci USA 78:848–851

Simon MA, Kornberg TB, Bishop JM (1983) Three loci related to the *src* oncogene and tyrosine-specific protein kinase activity in *Drosophila*. Nature 302:837–839

Snyder MA, Bishop JM (1984) A mutation at the major phosphotyrosine in $pp60^{v\text{-}src}$ alters oncogenic potential. Virology 136:375–386

Stehelin D, Varmus HE, Bishop JM, Vogt PK (1976) DNA related to the transforming gene(s) of avian sarcoma viruses is present in normal avian DNA. Nature 260:170–173

Stephens RM, Rice NR, Hiebsch RR, Bose HR Jr, Gilden RV (1983) Nucleotide sequence of v-*rel*: the oncogene of reticuloendotheliosis virus. Proc Natl Acad Sci USA 80:6229–6233

Stroobant P, Waterfield MD (1984) Purification and properties of porcine platelet-derived growth factor. EMBO J 3:2963–2967

Sutrave P, Bonner TI, Rapp UR, Jansen HW, Patschinsky T, Bister K (1984) Nucleotide sequence of avian retroviral oncogene v-*mil*: homologue of murine retroviral oncogene v-*raf*. Nature 309:85–88

Sweet RW, Yokoyama S, Kamata T, Feramisco JR, Rosenberg M, Gross M (1984) The product of *ras* is a GTPase and the T24 oncogenic mutant is deficient in this activity. Nature 311:273–275

Tabin CJ, Bradley SM, Bargmann CI, Weinberg RA, Papageorge AG, Scolnick EM, Dhar R, Lowy DR, Chang EH (1982) Mechanism of activation of a human oncogene. Nature 300:143–149

Takeya T, Hanafusa H (1982) DNA sequence of the viral and cellular *src* gene of chickens. II. Comparison of the *src* genes of two strains of avian sarcoma virus and of the cellular homolog. J Virol 44:12–18

Takeya T, Hanafusa H (1983) Structure and sequence of the cellular gene homologous to the RSV *src* gene and the mechanism for generating the transforming virus. Cell 32:881–890

Taparowsky E, Suard Y, Fasano O, Shimizu K, Goldfarb M, Wigler M (1982) Activation of the T24 bladder carcinoma transforming gene is linked to a single amino acid change. Nature 300:762–765

Taparowsky E, Shimizu K, Goldfarb M, Wigler M (1983) Structure and activation of the human N-*ras* gene. Cell 34:581–586

Thompson CB, Challoner PB, Neiman PE, Groudine M (1985) Levels of c-*myc* oncogene mRNA are invariant throughout the cell cycle. Nature 314:363–366

Toda T, Uno I, Ishikawa T, Powers S, Kataoka T, Broak D, Cameron S, Broach J, Matsumoto K, Wigler M (1985) In yeast, *RAS* proteins are controlling elements of adenylate cyclase. Cell 40:27–36

Tsuchida N, Ryder T, Ohtsubo E (1982) Nucleotide sequence of the oncogene encoding the p21 transforming protein of Kirsten murine sarcoma virus. Science 217:937–939

Ullrich A, Coussens L, Hayflick JS, Dull TJ, Gray A, Tam AW, Lee J, Yarden Y, Libermann TA, Schlessinger J, Downward J, Mayes ELV, Whittle N, Waterfield MD, Seeburg PH (1984) Human epidermal growth factor receptor cDNA sequence and aberrant expression of the amplified gene in A431 epidermoid carcinoma cells. Nature 309:418–425

Van Beveren C, Galleshaw JA, Jonas V, Berns AJM, Doolittle RF, Donoghue DJ, Verma IM (1981a) Nucleotide sequence and formation of the transforming gene of a mouse sarcoma virus. Nature 289:258–262

Van Beveren C, van Straaten F, Galleshaw JA, Verma IM (1981b) Nucleotide sequence of the genome of a murine sarcoma virus. Cell 27:97–108

Van Beveren C, van Straaten F, Curran T, Müller R, Verma IM (1983) Analysis of FBJ-MuSV provirus and c-*fos* (mouse) gene reveals that viral and cellular *fos* gene products have different carboxy termini. Cell 32:1241–1255

Van Beveren C, Enami S, Curran T, Verma IM (1984) FBR murine osteosarcoma virus. II. Nucleotide sequence of the provirus reveals that the genome contains sequences acquired from two cellular genes. Virology 135:229–243

Varmus H, Swanstrom R (1982) Replication of retroviruses. In: Weiss R et al. (eds) Molecular Biology of Tumor Viruses, 2nd edn. Cold Spring Harbor Laboratory, Cold Spring Harbor, New York, pp 369–512

Verma IM (1983) Organization and structure of retrovirus genomes. In: Becker Y (ed) Replication of Viral and Cellular Genomes. Martinus Nijhoff, Boston, pp 275–313

Wang L-H, Feldman R, Shibuya M, Hanafusa H, Notter MFD, Balduzzi PC (1981) Genetic structure, transforming sequence, and gene product of avian sarcoma virus UR1. J Virol 40:258–267

Waterfield MD, Scrace GT, Whittle N, Stroobant P, Johnsson A, Wasteson A, Westermark B, Heldin C-H, Huang JS, Deuel TF (1983) Platelet-derived growth factor is structurally related to the putative transforming protein $p28^{sis}$ of simian sarcoma virus. Nature 304:35–39

Watson DK, Reddy EP, Duesberg PH, Papas TS (1983) Nucleotide sequence analysis of the chicken c-*myc* gene reveals homologous and unique coding regions by comparison with the transforming gene of avian myelocytomatosis virus MC29, *gag-myc*. Biochemistry Proc Natl Acad Sci 80:2146–2150

Weinberg RA (1983) A molecular basis of cancer. Scient Amer 249(5):126–142

Wierenga RK, Hol WGJ (1983) Predicted nucleotide-binding properties of p21 protein and its cancer-associated variant. Nature 302:842–844

Wilhelmsen KC, Eggleton K, Temin HM (1984) Nucleic acid sequences of the oncogene v-*rel* in reticuloendotheliosis virus strain T and its cellular homology, the proto-oncogene c-*rel*. J Virol 52:172–182

Willingham MC, Pastan I, Shih TY, Scolnick EM (1980) Localization of the *src* gene product of the Harvey strain of MSV to plasma membrane of transformed cells by electron microscopic immunocytochemistry. Cell 19:1005–1014

Willumsen BM, Christensen A, Hubbert NL, Papageorge AG, Lowy DR (1984a) The p21 *ras* C-terminus is required for transformation and membrane association. Nature 310:583–586

Willumsen BM, Norris K, Papageorge AG, Hubbert NL, Lowy DR (1984b) Harvey murine sarcoma virus p21 *ras* protein: biological and biochemical significance of the cysteine nearest the carboxy terminus. EMBO J 3:2581–2585

Xu Y-H, Ishii S, Clark AJL, Sullivan M, Wilson RK, Ma DP, Roe BA, Merlino GT, Pastan I (1984) Human epidermal growth factor receptor cDNA is homologous to a variety of RNAs overproduced in A431 carcinoma cells. Nature 309:806–810

Yamamoto T, Nishida T, Miyajima N, Kawai S, Ooi T, Toyoshima K (1983) The *erb*-B gene of avian erythroblastosis virus is a member of the *src* gene family. Cell 35:71–78

Yasuda S, Furuichi M, Soeda E (1984) An altered DNA sequence encompassing the *ras* gene of Harvey murine sarcoma virus. Nucl Acids Res 12:5583–5588

Young HA, Shih TY, Scolnick EM, Rasheed S, Gardner MB (1979) Different rat-derived transforming retroviruses code for an immunologically related intracellular phosphoprotein. Proc Natl Acad Sci USA 76:3523–3527

Young HA, Rasheed S, Sowder R, Benton CV, Henderson LE (1981) Rat sarcoma virus: further analysis of individual viral isolates and the gene product. J Virol 38:286–293

Yuasa Y, Srivastava SK, Dunn CY, Rhim JS, Reddy EP, Aaronson SA (1983) Acquisition of transforming properties by alternative point mutations within c-*bas/has* human proto-oncogene. Nature 303:775–779